IMMUNOLOGY AND IMMUNE SYSTEM DISORDERS

MOLECULAR ASPECTS OF INFECTIOUS DISEASE

IMMUNOLOGY AND IMMUNE SYSTEM DISORDERS

IMMUNOLOGY AND IMMUNE SYSTEM DISORDERS

MOLECULAR ASPECTS OF INFECTIOUS DISEASE

MAZEN T. SALEH
EDITOR

Nova Science Publishers, Inc.
New York

For permission to use material from this book please contact us:
Telephone 631-231-7269; Fax 631-231-8175
Web Site: http://www.novapublishers.com

Library of Congress Cataloging-in-Publication Data

Molecular aspects of infectious diseases / editor, Mazen T. Saleh.
p. ; cm.
Includes bibliographical references and index.
ISBN 978-1-61728-690-2 (hardcover)
1. Molecular microbiology. 2. Communicable diseases--Molecular aspects.
3. Pathology, Molecular. I. Saleh, Mazen T.
[DNLM: 1. Communicable Diseases--genetics. 2. Communicable
Diseases--pathology. WC 100]
QR74.M625 2010
616.9'0471--dc22
2010027265

Published by Nova Science Publishers, Inc. † New York

Contents

In: Molecular Aspects of Infectious Diseases
Editor: Mazen T. Saleh
ISBN: 978-1-61728-690-2

Chapter I

Overview

Mazen Saleh
Department of Biology and Northen Ontario School of Medicine, Laurentian University, Sudbury, Ontario, Canada, P3E 2C6

The expression "infectious diseases" is a loosely used expression to describe diseases that are readily transmitted between hosts. It is therefore often used interchangeably with transmissible diseases. As a field, infectious diseases is the study of "infections". An infection is the successful detrimental colonization of a host, keeping in mind those humans, animals, and plants have their normal microflora that colonizes various external and internal structures. When the immune system of the host is compromised or the protective barriers of host that keep these microbes at bay are breached, certain species may cause a detrimental infection in the host. In this case, the infectious agent is referred to as an opportunistic pathogen. The term pathogen itself was traditionally used to refer to bacteria when discussing infections. This now has changed and is used to refer to any biological agent that causes pathology in the host as a result of an infection. This includes bacteria, viruses, protozoa, and parasitic worms; the study of the latter two forms is traditionally reserved for the field of parasitology. Infectious diseases also include the study of other acellular forms of infectious agents such as prion proteins and viroids (catalytic ribonucleic acid molecules).

An infection must be distinguished from intoxication. An infection results from the active growth of the infectious agent within the host. This is also true for prion protein. Although it is a protein and is therefore not capable of growth and multiplication, it is capable of "increasing" its number following the infection and causing various pathologies. Intoxication on the other hand occurs as a result of intake by the host of material contaminated with microbial toxins.

Infectious agents can be transmitted from the environment to the host or from host-to-host through several vehicles. There is direct host-to-host transmission through aerosols released during coughing and sneezing in respiratory and other infections, or indirectly through inanimate objects where body fluids may last for a sufficient period of time to transmit the agent. In these cases, direct contact between hosts is not required and is a condition that merits quarantine of subjects. The other direct mode of transmission requires contact between hosts and covers sexually transmitted diseases. The last mode of transmission is the vector-based transmission where large animals such as foxes, dogs, and rodents as well as insects such as mosquitos and flies transmit the infectious agent from host-to-host or from object-to-host.

Epidemiology

In the field of infectious diseases, epidemiology plays a central role in tracking and developing plans to contain an epidemic or an emerging infectious disease. In other words, epidemiologists focus on the source and the dynamics of transmission of an infectious disease. Epidemiologists will gather descriptive, analytical, and experimental methodology to provide a comprehensive picture of the source and the dynamics of disease transmission. This information is very valuable when devising plans to contain an infectious disease.

Epidemiologists use a number of terms in the field to describe various aspects of the disease transmission. A disease that has a steady incidence level in a geographic region is referred to as endemic in that region. When the incidence level of a disease increases above the expected or the basal level, it is referred to as an epidemic. When this phenomenon is observed worldwide, it is referred to as a pandemic. A sporadic disease is used to describe a disease that occurs unpredictably in one or several parts of a region but not at a level to threaten the health of the general public. It is also important to define such terms as incidence and prevalence. Incidence can be defined as the number of new cases for a disease over a given period of time and within a specific geographic location. Prevalence on the other hand is the total number of cases for a disease over a given period of time and within a specific geographic location.

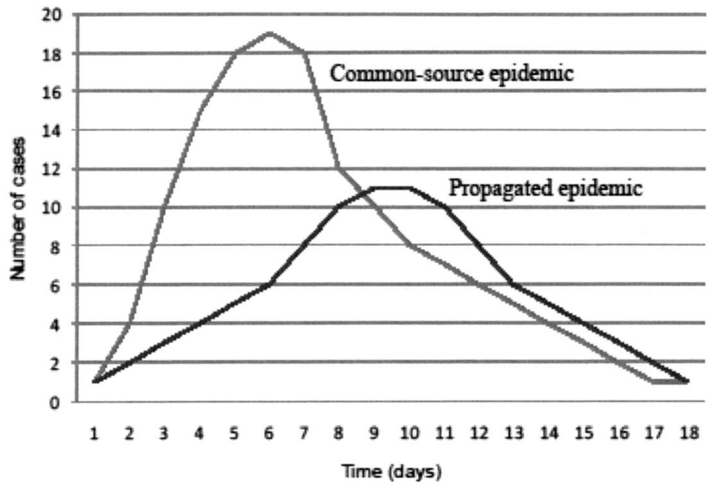

Figure 1.1. A simulated graph representing the two forms of epidemics. The common source epidemic is shown as an asymmetrical curve with a rapid increase in the number of cases for a given disease followed by a more gradual decrease. The propagated epidemic differs in that is symmetrical and also more gradual in the increase and decrease in the incidence of the disease.

The profile of an epidemic may take on one of two forms when the number of cases for the disease is monitored over time. The first form is a sharp increase in the number of cases followed by a more gradual decrease in the number of cases, thus producing an asymmetrical curve as shown in Figure 1.1. This is characteristic of the so-called common (or point) source epidemic. This situation arises when individuals are infected from a contaminated source of water or food poisoning. The second form displays a typical normal distribution curve symmetrical but more gradual than the common source form. This is characteristic of a disease where the infection spreads through person-to-person contact and is referred to as a propagated epidemic.

Host-Pathogen Interaction

Understanding the specifics in the interaction between pathogens and their hosts is crucial to the continuation of our effective intervention in containing and eradicating infectious diseases. These interactions are governed by the structure and physiology of the pathogens as well as by the structure and physiology of the host. The containment and eradication of a disease depends on the effectiveness of three key strategies we use in our clinics and hospitals: effective diagnostics, effective antimicrobials, and effective vaccines. Presently, we are also witnessing the emergence of a fourth strategy and that is the development of effective host response modulators, as in using interleukins, interferons, and inhibitors of intracellular signaling targets.

Recent developments in the field of molecular biology have exponentially increased our knowledge of host-pathogen interactions. The advent of gene cloning, genome sequencing, and bioinformatics, among other developments, has simplified these interactions from organismal level to molecular level. One may consider that host-pathogen interactions are mediated by specific molecular interactions between pathogen-derived and host-derived factors. Studying these interactions may be further facilitated by breaking down these interactions into discrete stages. The initial stage is the exposure of the host to the pathogen, followed by pathogen adhesion to specific surface receptors of specific tissue(s), and finally by host and pathogen responses. The last stage is the most complex and involves both extracellular and intracellular responses. The role of the host in these responses is being afforded more and more attention as we come to recognize the complexity of the host response even with the human genome sequence at hand.

To facilitate the understanding of microbial pathogenesis, the two major components in host-pathogen interactions are discussed. These are the role of the pathogen, in its surface structure and secreted virulence factors, and the role of the innate immune response of the host. Microbial surface structures contain the adhesins and ligands of various host cell surface receptors and thus mediate the initial interaction between the pathogen and the host. Adhesins and ligands may extend from the cytoplasmic membrane of the pathogen, or may be present in any sub-compartment exterior to the cytoplasmic membrane. These include molecules anchored to the peptidoglycan, the outer membrane (if present), slime layers, or appendages such as fimbriae and pili. These molecules are mostly proteins, glycoproteins, or glycolipids. The receptors on the surface of host cells involved in these interactions recognize specific molecular patterns (specific ligand-receptor complex) or non-specific molecular patterns called pathogen-associated molecular patterns (PAMPs). There are specific host cell receptors that recognize PAPMs and most are known as the toll-like receptors (TLRs). Other receptors are more specific in nature and include the scavenger receptors, mannose receptors, complement receptors, Fc receptors, and others.

A crucial microbial process for the successful establishment of infections follows adhesion and it is the elaboration of virulence factors and exotoxins. Virulence factor is a general term used to describe any microbial product required for the infection and/or the survival of the pathogen within the host. Exotoxins refer to proteins secreted by the pathogen with specific toxic activity. This toxicity does not have to be a lethal action against host cells and tissues but may simply be an enzyme that attenuates the physiology of the host. Some of the best characterized exotoxins include the tetanus toxin, diphtheria toxin, anthrax toxin, botulinum toxin, pertussis toxin, shiga toxin, *pseudomonas* exotoxin A, cholera toxin, and others.

Emerging Infectious Diseases

An emerging infectious disease (EID) refers to a disease that was not recognized previously or a previously recognized infectious disease but is presently causing an increase in the incidence of the disease. These EIDs include prion-related illnesses, severe acquired respiratory syndrome (SARS), swine flu, bird flu, and others. There are several reasons why a disease becomes an EID. The first is changes in the geography of human and animal habitat. These changes may come about as a result of encroachment of territory previously uninhabited by humans. These same changes may also come about as a result of global warming. Changes in the profile of pathogens and vectors that carry pathogens within a geographic zone are produced as global warming changes temperature distribution and water availability within these zones. For example, extension of certain shrubs from the United States into Canada will change insect populations accordingly. The second reason for the emergence of infectious diseases is the natural or xenobiotic-mediated mutations and genetic recombinations in pathogens. These mutations create new strains that allow pathogens to "jump hosts". This phenomenon has been observed recently with such EIDs as SARS, bird flu, and swine flu. The third possible reason for the emergence of such diseases is drug resistance. This is observed predominantly in bacterial pathogens than in other microbes or in viruses. When patients are prescribed antibiotics they are typically instructed to take the medication over a period of 7-10 days. Physicians, other health workers and government agencies emphasize the importance of following these instructions to patients and provide a variety of awareness and education tools for the patients and their families. Premature termination of treatment can produce antibiotic-resistant pathogens and that creates additional burden on the patient and on the health system. Drug resistance is produced naturally in microbial populations but because of the absence of selective pressure, those mutants do not have an obvious survival advantage over the parent population. This is not the case during infections where the treatment is terminated prematurely. This action would have applied selective pressure to favor the growth of the mutants, thus producing a drug-resistant infection. Drug resistance is acquired by microbes not strictly through mutations but may also occur as a result of genetic recombinations and acquisition of resistance factors. The latter is a class of small circular fragments of DNA called plasmids that can be transferred from a donor bacterium to a recipient bacterium often of the same genus. These plasmids contain genes that confer resistance to the recipient bacterium. Some of the more serious drug resistant pathogens we face today include the multi drug-resistant *Mycobacterium tuberculosis* (MDR-TB) and methicillin-resistant *Staphylococcus aureus* (MRSA). Resistance to antibiotics may result from one of the following changes to the physiology of the pathogen: (1) release of enzymes that inactivate the antibiotic, as in the case of β-lactamase and penicillin; (2) mutation in the target of the antibiotic so as to make insensitive to the drug; (3) metabolic bypass of a specific metabolic pathway inhibited by the antibiotic and; (4) expression of efflux pumps in the membrane of the pathogen that remove the drug from inside of the cell. We continue to face emergence of drug resistance in pathogens and while we have some options to deal with most of them, such as for example having first line, second line, and third line antibiotics; multi drug-resistance presents us with a difficult situation. This is particularly true in respiratory infections including Tuberculosis.

In: Molecular Aspects of Infectious Diseases
Editor: Mazen T. Saleh

ISBN: 978-1-61728-690-2

Chapter II

Structure and Function of Bacterial and Fungal Cell Walls

Johanna DeLongchamp[1], Mazen Saleh[1,2], and Garry Ferroni[1,2]
[1]Department of Biology and [2]Northen Ontario School of Medicine, Laurentian University, Sudbury, Ontario, Canada, P3E 2C6

Bacterial Cell Walls

The cell wall is a fairly rigid layer that lies just outside the cytoplasmic membrane. This layer confers the characteristic shape of the microorganism in addition to protecting against osmotic lysis. Several components contribute to the integrity and function of the cell wall thus affecting the penetrability and vulnerability of the cell (Costerton, In Gram and Cheng 1974). The shape and strength of the cell wall is mostly a result of the large polymer peptidoglycan (PG) also known as murein. Indeed, PG isolated from a cell retains the shape of the bacterium from which it was derived (Costerton, In Gram and Cheng 1974). PG is only found in *Bacteria* as the cell walls of *Archaea* and *Eukarya* lack either *N*-acetylmuramic acid, diaminopimelic acid, or both. Components important to PG synthesis.

Peptidoglycan contains two sugar derivatives, *N*-acetylglucosamine (NAG) and *N*-acetylmuramic acid (NAM), as well as four of five possible amino acids; L-alanine, D-alanine, D-glutamic acid, L-lysine and diaminopimelic acid (DAP). These subunits are connected to form repeating structures called glycan tetrapeptides or disaccharide-peptides (DSP), which in turn cross-link with each other to form a dense interconnected network (Prescott, Harley and Klein 2005). Cross links can be direct, most often between the carboxyl group of a terminal D-alanine and the amino group of a diaminopimelic acid, or indirect through a peptide interbridge. The PG backbone consists of alternating NAG and NAM sugars which are always connected through a β-1,4 linkage. The only major variation in the tetrapeptide chain is the lysine-DAP alternation, with only a few other select substitutions known. Currently there have been more than 100 PG types identified and the greatest source of variety occurs in the peptide interbridges which can be formed by any number of amino acids in addition to those found in the tetrapeptide (Madigan, Martinko and Parker 2003). Furthermore, there is much variety in the peptide cross linking itself as well as glycan chain length. For example cross linking has been reported to vary from 20-75% of the potential peptide cross links and glycan chains can be as long as 590 disaccharide-peptide units (DSP), as found in *Bacillus subtilis* (Shockman and Barrett 1983).

Considering the chain lengths and fairly extensive cross-linking, PG is a very large macromolecule and is in fact somewhat unique among biological macromolecules. Unlike nucleic acids, proteins or polysaccharides, PG is neither linear nor branched-linear but instead forms two and three-dimensional networks. PG also involves four different chemical linkages; glycosidic linkages between amino sugars, amide linkages between *N*-acetylmuramic acid and tetrapeptide side chains, peptide linkages between amino acids, and phosphodiester bonds that link accessory wall polymers with PG. Furthermore, the presence of D-amino acids results in L-D and D-L linkages (Shockman and Barrett 1983). The glycan strands have been hypothesized to have a vertical orientation, sticking up from the cell surface to form scaffolding. However, individual quantitative studies of *Escherichia coli* suggest that there is not enough PG to form such structures and instead propose that the glycan strands run horizontal to the cell surface, effectively wrapping around the circumference of the cell (Huang, et al. 2008).

Gram-Positive Cell Walls

The Gram-positive cell wall is made up mostly of a thick homogenous peptidoglycan layer which can range from 20–80 nm in thickness (Figure 2.1a). In addition to PG, Gram-positive bacteria cell walls contain cell-wall glycopolymers (CWG). Most Gram-positive species have at least two different CWGs in their cell wall, one associated with the PG and one associated with phospholipids of the cellular membrane. These CWGs can vary significantly with backbone sugars ranging from trioses to hexoses and net charges of oxidizing or reducing nature (Weidenmaier and Peschel 2008).

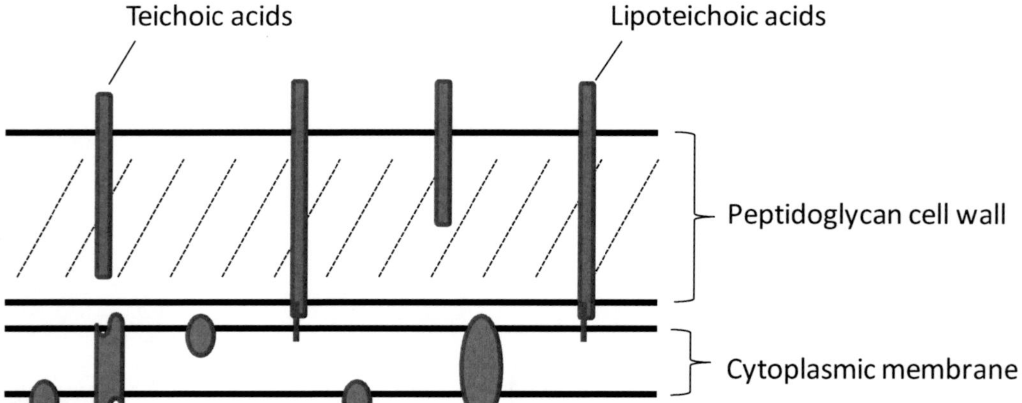

Figure 2.1 (a). Structure of the Gram-positive cell wall, depicting the thick peptidoglycan layer also containing wall-associated teichoic acids and membrane-associated lipoteichoic acids.

Acidic polysaccharides known as teichoic acids (TA) extend to the surface of the PG layer. These tightly bound sugars are often polymers of ribitol or glycerol phosphates, but may also be formed by tetroses, hexoses or complex sugars. TAs frequently contain additional sugars and amino acids such as D-alanine. Teichoic acids may also be linked to lipids in the cytoplasmic membrane and as such are referred to as *lipo*teichoic acids (LTA). LTAs demonstrate less diversity than PG associated TAs. Most teichoic acids exhibit zwitterionic properties due to the presence of negatively charged phosphate groups and modified free amino groups of attached amino acids (Weidenmaier and Peschel 2008). These acidic molecules contribute to the overall negative charge of the cell surface in addition to functioning in cell wall maintenance and enlargement during cell division (Madigan, Martinko and Parker 2003; Prescott, Harley and Klein 2005). It is proposed that the negative charge attracts and concentrates divalent cations at the cell surface, thus TAs may play a role in facilitating the passage of ions across the cell wall. It has also been demonstrated that the reactive groups of PG at the cell surface play a significant role in the metal-binding capabilities of Gram-positive cell walls (Shockman and Barrett 1983). NMR and X-ray studies suggest that the glycan strands of PG are relatively rigid and run parallel with each other, whereas the peptide bridges extend in all directions and acidic non-PG polymers are mobile and relatively flexible. Thus it is the mobility of these structures which allows expansion and contraction of the cell in response to environmental changes (Shockman and Barrett 1983).

In addition to the classical teichoic acids, many Gram-positive bacteria produce polyanionic CWGs. These polymers do not contain phosphate groups in their backbone and are often produced to replace PG-associated TA under conditions of limited phosphate. For example, *Bacillus subtilis* and other bacilli produce teichuronic acids, the anionic properties of which are a result of the presence of sugar derived acids known as uronic acids (Weidenmaier and Peschel 2008). Other secondary wall polysaccharides found among bacilli are anionic due to modification with pyruvyl groups. Furthermore, the cell walls of high G+C Gram-positive bacteria such as *Mycobacterium bovis* contain lipoglycans that are esterfied with anionic succinyl groups (Weidenmaier and Peschel 2008).

Initially scientists assumed a significant role for these CWGs in cell wall functions due to the universal distribution of the polymers among Gram-positive bacteria which of course coincides with the commitment of significant energy resources to the biosynthesis of the molecules. However with the creation of several defined CWG mutants, it became clear that many of these polymers are, more often than not, dispensable (under laboratory conditions) and likely play more of an indirect role within the cell wall. Proposed protective roles of CWGs include: anchoring outer protective layers such as capsules and slime layers, clogging cavities between PG strands, or modification of the physicochemical properties of the cell wall to hinder the entry of host defense molecules such as bacteriocins and antibiotics. Surface exposed CWGs impart the highly hydrophilic nature of the cell surface which in turn contributes to biofilm formation. Indeed, *Staphylococcus aureus* mutants with altered CWGs demonstrate significantly reduced ability to form biofilms and attenuation of virulence. Additionally, charged CWGs may contribute to the overall ionic environment of the cell wall as well as help maintain the proton gradient across the cytoplasmic membrane. One particular CWG found to be indispensable to cellular viability is lipoteichoic acid (LTA), which is suggested to be crucial in cell division. LTA-deficient *S. aureus* exhibits distorted cell shape specifically at division sites, implying that LTAs aid in positioning cell-division machinery.

It has also been found that many surface exposed CWGs act in specific receptor-mediated host interactions and immune responses. As such, these cell wall polymers are important pathogen associated molecular patterns (PAMPs) in innate immunity. More specifically, the majority of membrane associated CWGs have been shown to activate toll like receptor 2 (TLR2). CWGs also play a role in adaptive immunity. Various findings suggest that zwitterionic CWGs are important to T-cell dependent immune response and several antibodies against teichoic acids, as well as peptidoglycan, have been detected. It follows then, that CWGs may also play a role in evasion of the host immune response through antigenic variation (Weidenmaier and Peschel 2008).

Mycobacterial Cell Walls

Mycobacterial species are classified as high G+C Gram positive bacteria but exhibit unique characteristics in their cell walls that differentiate them from other Gram-positive species. For example, the peptidoglycan of mycobacteria contains N-glycolylmuramic acid instead of the aforementioned N-acetylmuramic acid. The glycol groups of the muramic acid residues may allow for a more compact packing of the peptidoglycan which could lead to increased opportunities for hydrogen bonding (Brennan and Nikaido 1995). An arabinogalactan layer is also found exterior, and connected via phosphodiester bonds, to the peptidoglycan layer. However the most distinctive element of the mycobacterial cell wall is the presence of mycolic acids. These lipids, consisting of branched fatty acid chains 60 to 90 carbons in length, make up approximately 60% of the weight of the cell wall. About 10% of arabinose residues in the arabinogalactan layer are in fact substituted by covalently linked mycolic acids (Jarlier and Nikaido 1994). Mycobacterial cell walls also contain several proteins and so-called "extractable lipids" which are non-covalently linked to the peptidoglycan-arabinogalactan skeleton. These lipids include trehalose-containing glycolipids, phenol-phthiocerol glycosides and glycopeptidolipids. These components assemble into an asymmetric bilayer of considerable thickness (Figure 2.1b) (Brennan and Nikaido 1995). Mycolic acids contain very few double bonds or cyclopropane groups which results in a very low fluidity of the lipid layer. As such, it is suggested that this reduced fluidity leads to reduced permeability of lipophilic solutes, effectively resulting in a permeation barrier.

The permeability of hydrophilic solutes was experimentally measured in *M. chelonae* as three orders of magnitude less than the outer membrane of *Escherichia coli* and ten orders of magnitude less than the *Pseudomonas aeruginosa* outer membrane (Brennan and Nikaido 1995). There is variation in the degree of permeability among the Mycobacteria species, of up to a factor of 100, which is directly correlated with the variations of cell wall components and organization (Jarlier and Nikaido 1994). Mycobacteria can be categorized into two groups: slow growers and fast growers. Interestingly, fast growers have higher hydrophilic permeability when compared to slow growers. This is likely due to the fact that a higher hydrophilic permeability also means a more efficient acquisition of hydrophilic nutrients such as glycerol and glucose (Brennan and Nikaido 1995). It is suggested that permeation of hydrophilic solutes occurs via aqueous channels in the cell wall. Porins have in fact been found in mycobacterial cell walls but are a minor protein component and possess significantly lower permeability than those of *E. coli*.

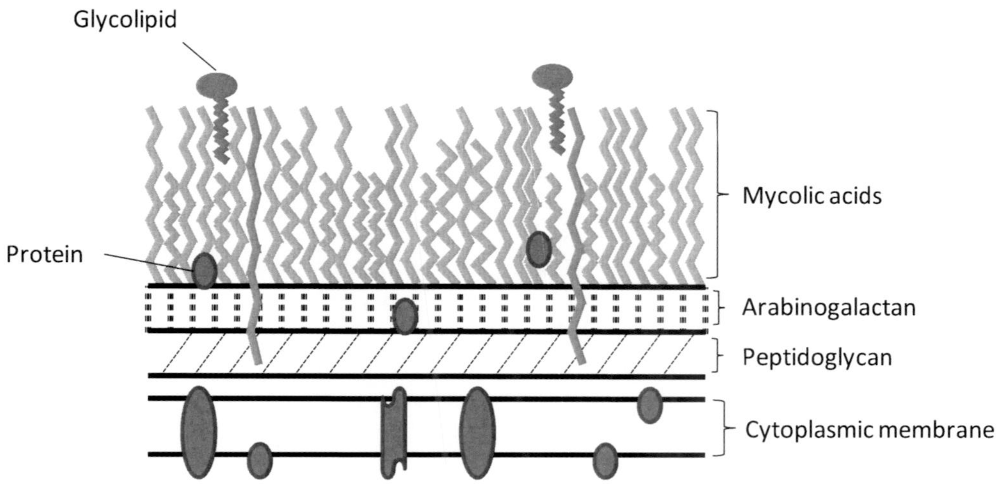

Figure 2.1 (b). Structure of the Mycobacterial cell wall containing long mycolic acids covalently linked with an arabinogalactan layer which is in turn connected via phosphodiester bonds to the peptidoglycan. Closely associated with the mycolic acids are proteins as well as bound and extractable lipids.

Lipophilic solutes, however, are not favored solutes for the porin pathway and must instead transverse the lipid bilayer. As mentioned above, the reduced fluidity of the pseudo-outer membrane composed mainly of long-chained and tightly packed mycolic acids, results in decreased permeability of these hydrophobic solutes. Nonetheless, diffusion of lipophilic solutes across the bilayer remains an important pathway and studies of lipophilic antibacterial agents show that an increase in hydrophobicity leads to an increase in efficacy of the drug (Brennan and Nikaido 1995).

This reduced permeability of the mycobacterial cell wall leads to an increased resistance to chemical injury, dehydration and antibiotics which in turn results in persistence of infection. Certain cell surface lipids of *Mycobacteria* may aid in the hallmark impairment of phagosome-lysosome fusion and exert endotoxin-like effects. These lipids, which may be readily released from the cell wall, have been implicated in both inflammatory and immunosuppressive responses. Of course, similar to the cell wall glycopolymers common to most Gram-positive bacteria, mycolic acids and other surface lipids of mycobacteria are also important pathogen-associated molecular markers for innate and adaptive immunity (Rhoades and Ullrich 2000).

Gram-Negative Cell Walls

If we take the cell wall to include the peptidoglycan layer, periplasm and outer membrane (Figure 2.2), it is clear the cell wall of Gram-negative species is far more complicated than that of Gram-positive organisms. In contrast with the thick Gram positive cell wall composed mainly of PG, Gram negative bacteria have only a thin 1-3nm PG layer, often comprised of only a couple sheets of PG. Another distinct difference between the two PG layers is the open molecular structure of Gram-negative PG. The polysaccharide chains are fairly dispersed and there is comparatively little cross linking between peptide side chains suggesting that, unlike Gram positive cell walls, the PG of Gram-negative bacteria does not serve as much of a permeation barrier to small molecules (Costerton, Ingram and Cheng 1974). As mentioned before, PG offers rigidity and protection; however the thinness of the Gram negative cell wall offers greater flexibility (Prescott, Harley and Klein 2005). In Gram-negative bacteria PG seems to play several indirect roles and if disrupted, affects outer membrane functions such as flagella assembly (Costerton, Ingram and Cheng 1974).

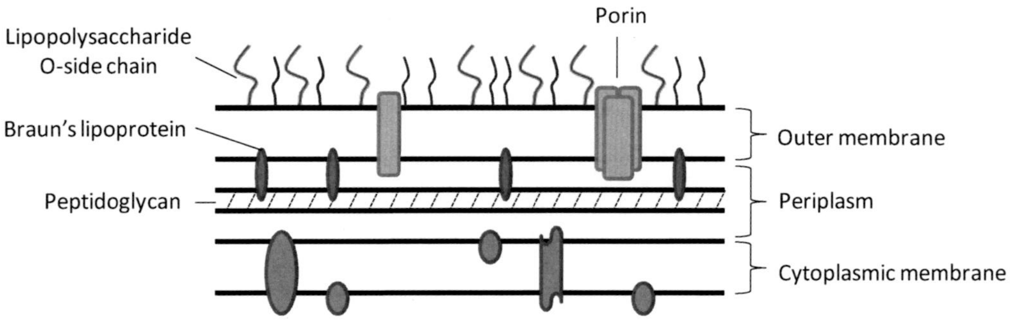

Figure 2.2. Structure of the Gram-negative cell wall including the peptidoglycan layer, periplasmic space and outer membrane. The external leaflet of the outer membrane is predominately lipopolysaccharides. The outer membrane and peptidoglycan are linked together by Braun's lipoprotein.

The defining feature of the Gram-negative cell envelope is the lipopolysaccharide (LPS) containing outer membrane. This outer barrier, often referred to as a 'molecular sieve', is responsible for regulating the flux of molecules into and out of the periplasmic space. This filtering mechanism limits the number of molecules bombarding the inner membrane while keeping key enzymes within the periplasm. Perhaps most important is the role of the outer membrane in restricting access of antibiotics. The outer membrane and peptidoglycan layer are firmly linked together via Braun's lipoprotein which is covalently linked with the PG on one end while the other end hydrophobically interacts with the lipids of the outer membrane. It is well known that this anchoring by Braun's lipoprotein is strong enough to allow the PG layer and outer membrane to be experimentally isolated as one unit. In addition there have been several adhesion sites identified between the cytoplasmic membrane and peptidoglycan layer, this association is likely also due to a shared linkage with a membrane protein (Prescott, Harley and Klein 2005).

Unlike the cytoplasmic membrane, the outer membrane is asymmetrical with phospholipids on the inner leaflet and a mixture of phospholipids and LPS on the outer leaflet. Membrane proteins of the outer membrane also differ significantly from inner membrane proteins. Integral proteins that span the cytoplasmic membrane are formed by hydrophobic α-helices. In contrast, the outer membrane proteins are formed by antiparallel amphipathic β-strands that fold into β-barrels with a hydrophilic interior. In this case, the side chains of the hydrophobic residues extend outwards to make contact with membrane lipids (Bos, Robert and Tommassen 2007). Since the outer membrane is not energized by a proton gradient and the periplasmic space does not contain ATP, the majority of transmembrane transport is achieved by means of passive diffusion through water filled outer membrane porins (OMPs). These trimeric channels allow the passage of hydrophilic solutes up to

600Da. It should be noted however that energy coupling complexes which harness energy from the inner membrane for processes at the outer membrane have been identified (Bos, Robert and Tommassen 2007).

As mentioned earlier, LPS is the most distinctive component of the Gram-negative cell wall. These complex molecules are formed of three parts: lipid A, a core polysaccharide and the O side chain. The lipid A portion obviously anchors the LPS in the outer membrane while the remainder of the molecule projects from the surface of the cell. The O side chain is a polysaccharide chain which extends outward from the core sugar and is also referred to as the O antigen because it functions as a receptor and is readily recognized by host antibodies. It follows then that bacteria may use this recognizable feature to deceive the host immune response by rapidly and frequently changing the composition of the side chain to avoid detection. Similar to the negative charge imparted by teichoic acids in Gram-positive bacteria, LPS also contributes to the cell surface negative charge.

While LPS contributes structurally by stabilizing the outer membrane, it is also classified as an endotoxin due to the toxic lipid A portion of the molecule. Lipid A is the only part of LPS recognizable by the innate immune response, is highly immune stimulatory and may illicit a response even at low concentrations. LPS is primarily identified by the toll like receptor 4 (TLR4), with the help from the LPS binding protein (LBP), and activates a signaling cascade which ultimately results in the activation of NFκB and the pro-inflammatory response (Samuel, Ernst and Bader 2005).

Fungal Cell Walls

The cell wall of fungi is a very dynamic structure subjected to several changes above and beyond regular cell growth and division including spore germination, hyphal branching and septum formation in filamentous fungi. The maintenance of cell wall plasticity is likely due to the hydrolytic enzymes (chitinases, glucanases and peptidases) found intimately associated with cell wall components (Adams 2004). The complexity and size of different fungi varies but most are enclosed in a cell wall of chitin extensively cross-linked with several polysaccharides, most commonly glucans and mannans, as well as largely modified glycoproteins (Figure 2.3).

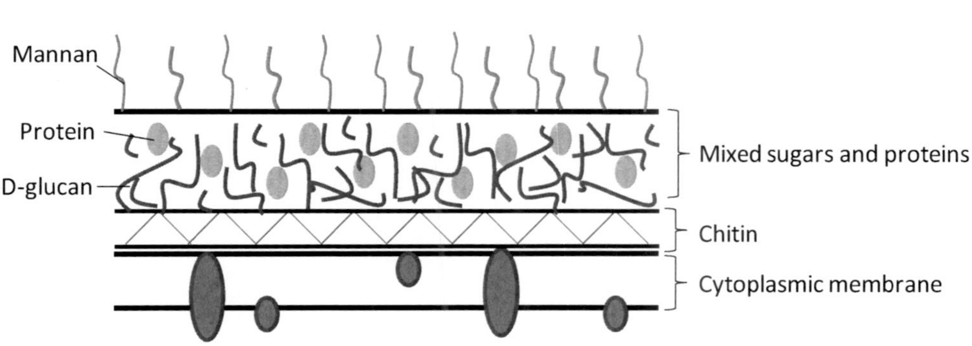

Figure 2.3. The cell wall of fungi. An envelope of chitin cross-linked with polysaccharides such as glucan and mannan, extensively modified glycoproteins and several enzymes. The surface exposed sugar polymers, often mannans, act as antigenic markers.

While structurally important, chitin is considered a relatively minor constituent of the fungal cell wall, making up only 1-2% of the yeast cell wall by dry weight and 10-20% of filamentous fungal species such as *Aspergillus* (Bowman and Free 2006). Chitin is composed of long chains of β-1,4-linked N-acetylglucosamine (NAM) residues which, through the formation of inter-chain hydrogen bonds and subsequent crystallization in the extracellular space, form exceptionally strong chitin microfibrils.

Glucans are the major structural sugars of the cell wall, making up 50-60% of the total cell wall by dry weight. The glucan polymers are primarily chains of β-1,3-linked glucose molecules but several other linkages have been identified such as β-1, 6-glucan. As the major structural constituent, glucans are required for proper cell function and anchor other cell wall components through covalent linkages. Glucan chains are often as long as 1500 glucose residues with 40-50 of these residues becoming sites of attachment for further 1,3-glucan chains, resulting in a branched structure.

Tightly interwoven with the chitin and glucan components are the glycoproteins modified with N- and O-linked polysaccharides. Fungal cell walls commonly contain proteins glycosylated with mannose rich chains, known as mannoproteins. Other glycoproteins found in the cell wall may be modified with other sugars such as galactose or combinations of sugars. The outermost layer of the fungal cell wall is often a polysaccharide coating with variation in the density of the sugar polymers among species. It is proposed that it is this layer that is largely responsible for the permeability of the microorganism (Arana, et al. 2009). Some cell wall proteins have also been found to be anchored directly with the plasma membrane and cell wall through a glycosylphosphatidylinositol (GPI) foundation. Non GPI-anchored proteins are integrated into the cell wall via covalent linkages between their sugar side branches and those of chitin and glucans. Yeast cell walls contain 30-50% protein while those of filamentous fungi contain 20-30% and interestingly, some of these proteins are actually 'intracellular' proteins. Once believed to be contaminants of fungal cell wall preparations, it is now understood that these cytosolic and mitochondrial proteins are indeed cell wall constituents, though their purpose is unclear (Bowman and Free 2006). Cell wall proteins can be classified as either structural components that lack catalytic domains, or wall-associated enzymes which exhibit catalytic activity and are mostly involved with the synthesis and remodeling of the cell wall. As described above, the hydrolytic enzymes such as chitinases, glucanases and peptidases are involved in breaking bonds and breaking down wall components while synthesis enzymes such as glycosyltranferases are responsible for the formation of new bonds and cross-linking between wall polymers. That is not to say that the structural proteins do not play a role outside of scaffolding, many of these proteins aid in cell migration, adhesion, fusion and mating (Bowman and Free 2006).

Different sugar polymers found on the fungal cell wall, O-linked mannan for example, are identifiable by several toll-like receptors (TLRs). In addition, several of the fungal polysaccharides are recognized by immune cell receptors such as the C-type lectin receptor (CLR) and the macrophage mannose receptor (MR). Consequently, recognition by these receptors ultimately leads to an immune response in efforts to eradicate the pathogen.

Conclusion

Clearly cell walls, as the outermost structure of these microorganisms, are primarily responsible for mediating interactions with the host. The cell wall is involved in adhesion, colonization, signaling and immune recognition, making it a very important player in infection. In addition to significant host-microorganism interactions, the cell wall is essential for survival of the microorganism. As such, the cell wall makes a great target for drug discovery in antimicrobials and antifungals. Indeed many of the antibiotics on market today act against microbial cell walls. Penicillin, the first antibiotic applied on a large scale, inhibits cell wall biosynthesis enzymes by means of its beta-lactam ring. In the years since penicillin's introduction, hundreds of beta-lactam based antibiotics have been introduced. Additional classes of antibiotics that attack the same biosynthesis pathway include glycopeptide-, lipopeptide-, and lipodepsipeptide-antibiotics, as well as lantibiotics and several natural products (Schneider and Sahl 2010). Research and discovery into antifungal agents however, significantly lags behind that of antibacterials for several reasons. Disease-causing fungi were not recognized as serious pathogens until recently. Notably, the rate, severity, and diversity of fungal infections have greatly increased in the past ten years (Arana, et al. 2009). It wasn't until the introduction of imidazoles and triazoles in the late 1980's and early 1990's that we were able to safely treat systemic fungal infections. These azole compounds target the formation and function of ergosterol, an important component of fungal cell membranes (Ghannoum and Rice 1999). It took another 15 years before the first antifungal to effectively target the cell wall was released. In 2002 echinocandins, which inhibit the synthesis of β-D-glucan, were introduced (Denning 2002).

References

Adams, David J. "Fungal cell wall chitinases and glucanases." *Microbiology* 150 (2004): 2029-35.

Arana, David M, Daniel Prieto, Elvira Roman, Cesar Nombela, Rebeca Alonso-Monge, and Jesus Pla. "The role of the cell wall in fungal pathogenesis." *Microbial Biotechnology* 2, no. 3 (2009): 308-20.

Bos, Martine P, Viviane Robert, and Jan Tommassen. "Biogenesis of the Gram-negative bacterial outer membrane." *Annual Review of Microbiology* 61 (2007): 191-214.

Bowman, Shaun M, and Stephen J Free. "The structure and synthesis of the fungal cell wall." *BioEssays* 28, no. 8 (2006): 799-808.

Brennan, Patrick J, and Hiroshi Nikaido. "The envelope of mycobacteria." *Annual Review of Biochemistry* 64 (1995): 29-63.

Costerton, J W, J M Ingram, and K J Cheng. "Structure and function of the cell envelope of Gram negative bacteria." *Bacteriological Reviews* 38, no. 1 (1974): 87-110.

Denning, David W. "Echinocandins: a new class of antifungal." *Journal of Antimicrobial Chemotherapy* 49 (2002): 889-91.

Ghannoum, Mahmoud A, and Louis B Rice. "Antifungal agents: Mode of action, mechanisms of resistance and correlation of these mechanisms with bacerial resistance." *Clinical Microbiology Reviews* 12, no. 4 (1999): 501-17.

Huang, Kerwyn Casey, Ranjan Mukhopadhyay, Bingni Wen, Zemer Gitai, and Ned S Wingreen. "Cell shape and cell-wall organization in Gram-negative bacteria." Edited by 87. *PNAS* 105, no. 49 (2008): 19282.

Jarlier, Vincent, and Hiroshi Nikaido. "Mycobacterial cell wall: Structure and role in natural resistance to antibiotics." *FEMS Microbiology Letters* 123 (1994): 11-18.

Madigan, Michael T, John M Martinko, and Jack Parker. *Brock Biology of Microorganisms.* 10th Edition. Upper Sadle River: Prentice Hall, 2003.

Prescott, Lansing M, John P Harley, and Donald A Klein. *Microbiology.* 6th Edition. New York: McGraw Hill, 2005.

Rhoades, E R, and H J Ullrich. "How to establish a lasting relationship with your host: Lessons learned from Mycobacterium spp." *Immunology and Cell Biology* 78 (2000): 301-10.

Samuel, Miller I, Robert K Ernst, and Martin W Bader. "LPS, TLR4 and infectious disease diversity." *Nature Reviews Microbiology* 3 (2005): 36-47.

Schneider, Tanja, and Hans-Georg Sahl. "An oldie but a goodie - cell wall biosynthesis as antibiotic target pathway." *International Journal of Medical Microbiology* 300, no. 2-3 (2010): 161-169.

Shockman, Gerald D, and John F Barrett. "Structure, Function and Assembly of Cell Walls of Gram-positive Bacteria." *Annual Review Microbiology* 37 (1983): 501-27.

Weidenmaier, Christopher, and Andreas Peschel. "Teichoic acids and related cell-wall glycopolymers in Gram-positive physiology and host interactions." *Nature Reviews Microbiology* 6 (2008): 276-87.

In: Molecular Aspects of Infectious Diseases
Editor: Mazen T. Saleh

ISBN: 978-1-61728-690-2

Chapter III

Bacterial Cell Surface Structures Important in Pathogenesis

***G. D. Ferroni*[1,2]*, L.G. Leduc*[1]*, and N.C.S. Mykytczuk*[1]**

Department of Biology, Laurentian University, Sudbury, Ontario, Canada P3E 2C6[1]

Northen Ontario School of Medicine, Laurentian University, Sudbury, Ontario, Canada P3E 2C6[2]

Importance of the Bacterial Cell Surface

The structures found at the surface of a bacterial cell are important to the cell's survival in the natural environment. Cellular surface structures such as the lipopolysaccharide (LPS), the peptidoglycan, fimbriae, pili, etc. allow the cell to communicate with both the living and nonliving components of the environment. Also, and most importantly, these structures allow for the cell's existence and survival in the environment. There is no doubt that bacterial cell surface structures have a fundamental biological function to the cell itself. However, such structures can also be important in the pathogenesis of infectious diseases. Indeed, bacterial cell surface structures may play various roles in pathogenesis such as providing permeability barriers that allow essential nutrients in and block antimicrobial substances out of the cell, permitting the bacterial cell to stick to the host cell by way of adhesion molecules, offering enzymes that catalyze chemical reactions necessary for the survival of the bacterium, providing protective barriers that protect the bacterial cell from the host's defense mechanisms, exposing endotoxins that cause inflammatory response in the host, and triggering signal transduction that ultimately expresses some determinant of virulence such as exotoxin. This chapter provides an overview of bacterial surface structures important in pathogenesis. The cellular constituents of the bacterial cell surface examined here are the outermost layers, i.e., capsules, slime layers, and S-layers.

Bacterial Capsules and Slime Layers

The capsules and slime layers of bacteria are external to the cell wall and are composed of substances often referred to as exopolymers. They are not covalently attached to the bacteria. Capsules are thick (relative to the other cell structures), viscous, halo-like structures that surround the cells of certain species. Slime layers are less distinct than capsules in that they lack a definite form and boundary. The exopolymers of both capsules and slime layers are usually polysaccharides, but there are a few *Bacillus* species that produce polypeptides. An example is *Bacillus anthracis* which has a capsule, during in vivo multiplication at least, of poly-D-glutamic acid. The term glycocalyx is also used to refer to polysaccharide-containing structures external to the bacterial cell wall, but it includes capsules, slime layers, and S layers (Costerton et al., 1978; Costerton et al., 1981). Because capsules and slime layers can be virulence factors for pathogenic bacteria and are important in the antigenic characterization of pathogens, they have been much studied.

A variation on capsules and slime layers is the biofilm, a network of extracellular polysaccharides in which are embedded microbial species that comprise a dynamic community. The biofilm is the major mechanism by which microbial attachment to surfaces occurs in natural environments. Thus, any implanted medical device can encourage biofilm formation, which has implications for the function of the implant and indeed the development of life-threatening infectious diseases. Moreover, residents of the biofilm, in contrast to their planktonic relatives, tend to be more resistant to antimicrobials (Gilbert et al., 1997) and to elimination by our immune systems (Costerton et al., 1999). Biofilms also play a role in infectious disease development in the host devoid of implants. As pointed out in a review of microbial biofilms by Davey and O'Toole (2000), this link of biofilms to pathogenesis is likely true for lung infections by *Pseudomonas aeruginosa* in cystic fibrosis patients, chronic otitis media, and periodontitis. The role of the biofilm in disease pathogenesis is also the subject of a review by Parsek and Singh (2003) which considers infectious kidney stones, bacterial endocarditis, and cystic fibrosis airway infections, in detail, as well as the biofilm as a reservoir for bacterial pathogens.

The presence of a capsule or slime layer influences the macroscopic appearance of colonies on solid media. Colonies of encapsulated bacteria are often viscous, a trait referred to as mucoidy. A rather extreme example of this is *Klebsiella pneumoniae* growing on blood agar. Colonies of non encapsulated bacteria of the same species, however, are smaller and lack the slimy, glistening appearance.

Capsules can be visualized using light microscopy and various staining procedures, both negative and direct. They can also be studied using the electron microscope and especially the confocal scanning laser microscope. The latter allows the visualization of hydrated samples and thus gives a better picture of three-dimensional structure. A serological procedure that demonstrates capsules and can also be used for bacterial identification is the Quellung reaction (Murray et al., 2007). The clinical sample and antiserum (antisera against a variety of capsular serotypes are commercially available) are mixed on a slide and viewed by phase-contrast microscopy. If an antigen-antibody reaction has occurred, the capsule will appear swollen and, therefore, more distinct.

Although fimbriae and pili are the bacterial structures most often involved in attachment, capsules and slime layers can mediate attachment to solid surfaces, including host cells. In other words, the binding of a bacterial pathogen to the specific receptor on a host cell can be due to the capsular polysaccharide.

It is generally accepted that capsules and slime layers can play a role in protecting the cells that elaborate them, against desiccation, phage attachment, and hydrophobic chemicals (Wilkinson, 1958). Resistance to desiccation follows from the fact that polysaccharide capsules tend to be extremely hydrated. Resistance to phage attachment is because receptor sites on the cell wall are harder to reach. Protection against hydrophobic chemicals such as detergents is due to the hydrophilicity of the extracellular polysaccharides.

In the case of pathogenic bacteria, the capsule plays a role in virulence by interfering with aspects of the host's system of innate immunity. Specifically, capsules can interfere with phagocytosis by host white blood cells (Allen et al., 1987; Cross, 1990), opsonization (Abeyta et al., 2003; O'Riordan and Lee, 2004), and complement activation (Pluschke and Achtman, 1984) and complement functions (Nicholson and Glynn, 1975; Moxon and Kroll, 1990), all of which are part of the inflammatory response induced by host invasion. The importance of capsules as a virulence factor is indicated by the response of mice to encapsulated and non encapsulated strains of *Streptococcus pneumoniae*. Whereas a few encapsulated cells will kill a mouse when injected intraperitoneally, thousands of non encapsulated cells must be injected to have the same effect.

Notable pathogens that produce antiphagocytic capsules are *Streptococcus pneumoniae* (the pneumococcus), *Streptococcus pyogenes, Staphylococcus aureus, Haemophilus influenzae*, *Neisseria meningitidis* (the meningococcus), *Klebsiella pneumoniae*, *Escherichia coli*, *Bacteroides fragilis*, and *Cryptococcus neoformans*. Included in this list are the main etiological agents of pneumonia and meningitis, and as pointed out by Finlay and Falkow (1989), that all possess polysaccharide capsules attests to the importance of capsules in the infective process.

Phagocytosis is the process by which certain white blood cells (neutrophils, monocytes, macrophages, and dendritic cells) recognize and destroy extracellular microbial pathogens. Phagocytosis can occur when a microbial surface ligand interacts with a phagocytic receptor. Moreover, phagocytosis can be enhanced if the microbial cell is first opsonized, that is coated by antibodies, a mannose binding lectin, or the glycoprotein C3b which is an intermediate in the process of complement activation. These opsonins provide a linkage between the microbial cell and the phagocyte.

The production of the opsonin C3b occurs via the complement system, a system involving numerous serum and membrane-bound proteins that can result in enhanced phagocytosis, an enhanced inflammatory response, and microbial cell lysis. Complement activation is triggered by an antigen-antibody complex on a microbial cell surface, a microbial surface macromolecule such as the lipopolysaccharide (LPS) found in the outer leaflet of the outer membrane of Gram-negative bacteria, or a mannose-binding lectin, and proceeds via one of three different pathways (determined by the trigger), all of which produce C3b. C3b proceeds in the pathway and eventually the "membrane attack complex" is produced that inserts into the bacterial membrane and brings about cell lysis. Moreover, the mere presence of C3b on the microbial surface enhances phagocytosis due to phagocytes possessing receptors for C3b.

A key question is how the presence of a capsule mitigates phagocytosis and complement activation, two processes that ultimately kill the microbe. The answer is that the blanketing of the microbial cell surface with hydrophilic polysaccharide prevents the microbial ligand-phagocyte receptor interactions necessary for opsonin-independent phagocytosis and opsonin-dependent phagocytosis (Abeyta et al., 2003), and the triggering of complement activation. With regard to the latter, the failure to produce C3b ensures that the "membrane attack complex" is not produced and that C3b-phagocyte association does not occur. The effectiveness of the capsule in insulating the microbial cell surface from complement, antibodies, etc. is related to the physical integrity of the capsule.

Nelson et al. (2007) have proposed another role for the capsule in the cycle of virulence. Unencapsulated cells of *Streptococcus pneumoniae* were shown to be less capable than encapsulated cells of mouse nasal colonization. The difference was not due to greater susceptibility to opsonophagocytosis on the part of the unencapsulated cells, but to their being more easily bound to nasal mucus. This made them less likely to transit to the epithelial surface where stable colonization occurs. Thus capsules may contribute to virulence by having an inhibitory effect on in vivo clearance through mucus binding.

Thus the capsule protects the cell against phagocytosis and interferes with the activation of complement. But, this interference can be lessened if the host produces antibody against the capsular polysaccharide, because antigen-antibody complexes are a trigger for one of the complement pathways. Not surprisingly, some important bacterial pathogens are able to counteract this process. *Streptococcus pneumoniae* and *Streptococcus pyogenes* synthesize a capsular polysaccharide of glycosaminoglycan hyaluronic acid, which is also a constituent of the extracellular matrix of animal tissue. Because the capsule is not immunogenic, the usual anticapsular antibody response does not occur and opsonization does not ensue. Some strains of *Neisseria meningitidis* elaborate a capsule that is largely a sialic acid polymer. Sialic acid is a component of some glycoproteins found in animal tissue, and thus the *N. meningitidis* capsule, for these strains at least, in non-immunogenic.

The immunogenicity of capsular polysaccharides has been exploited in the production of vaccines effective against *Haemophilus influenzae* type b, *Streptococcus pneumoniae*, and *Neisseria meningitidis* serogroup C. *H. influenzae* type b causes important primary infections in children aged five months to five years, including meningitis, epiglottitis, bacteremia, cellulitis, pneumonia and septic arthritis. Vaccines against this strain consist of polyribosylribitol phosphate capsular polysaccharide complexed with either tetanus protein or an outer membrane protein from *Neisseria meningitidis* serogroup B. Administering the vaccine by separate injection or in combination with diphtheria/pertussis/tetanus/polio immunization, according to a set regime, is highly protective.

Streptococcus pneumoniae is the most common cause of bacterial infection in children under the age of two years. The conditions of concern are pneumonia, bacteremia, meningitis, and otitis media, especially as the pneumococcus can be multiple-antibiotic resistant. Besides being common in this age group, invasive pneumococcal disease affects the elderly and certain high risk groups including those with functional or anatomic asplenia and those with immunodeficiencies such as acquired immune deficiency syndrome. Available vaccines include pneumococcal conjugate vaccines containing capsular antigens from seven *S. pneumoniae* serotypes each conjugated to a variant of the diphtheria toxin; and, pneumococcal polysaccharide vaccines each with capsular antigens from 23 different serotypes.

The most common *Neisseria meningitidis* serogroups implicated in invasive meningococcal disease, mainly meningitis and septicemia, are A, B, C, W-135, and Y. All but serogroup B contain immunogenic capsules and are thus potential immunizing agents.

The two main types of meningococcal vaccines are referred to as protein-polysaccharide conjugate vaccines and capsular polysaccharides vaccines. Available conjugate vaccines include group C oligosaccharides conjugated to a variant of the diphtheria toxin or to the tetanus toxoid. Available polysaccharide vaccines include groups A and C combined or groups A, C, Y and W-135 combined.

In summary, these structures of exopolymeric material manifest themselves as capsules or slime layers and are important to the producing cell as mediators of attachment or protection. In pathogenic microbes, the protective element is directed against phagocytosis, the activation of complement and aspects of complement activity. The antigenic nature of capsular polysaccharides has been and will continue to be exploited in the production of effective vaccines.

S-Layers

The outermost surface structure of both Gram-positive and Gram-negative Archaea and Bacteria consists of a monomolecular crystalline assemblage of protein subunits called surface layers or S-layers (Costerton et al., 1981; Sara & Sleytr, 2000; Sletyr, 1978). In other words, S-layers form a regular array of glycoprotein subunits at the surface of the bacterial cell, e.g. the ordered glycoproteins found on the surface of cells of *Spirillum volutans*.

Although S-layers seem to be ubiquitous in prokaryotes, they are particularly well documented in the Archaea. Indeed, almost all Archaea have S-layers. These surface structures have often gone unnoticed because they are usually lost after prolonged cultivation in the laboratory (Sara & Sleytr, 2000).

S-layers can have oblique, square or hexagonal symmetry, the latter predominating in Archaea. With respect to lattice type, one morphological unit consists of 1, 2, 3, 4 or 6 identical glycoproteins. The glycoprotein units show centre-to-centre spacing of 2.5 to 3.5 nm. The overall thickness of S-layers is in the range 5 to 25 nm with a smooth exterior surface and a rough or corrugated interior surface. In some cases (i.e. many S-layer lattices of Archaea), cylindrical-like extensions of the interior surface are present. As can be expected, pores (2 to 8 nm in diameter) occupy a good part of the total surface area of S-layers. In Gram-positive bacteria, S-layer subunits are attached to the peptidoglycan or the pseudopeptidoglycan (in archaeans). In Gram-negative bacteria, the subunits are bound to the lipopolysaccharides (LPS) of the outer membrane. In bacteria lacking a rigid cell wall, the S-layer subunits are linked to the cell membrane. As mentioned earlier, S-layers are easily lost from the surface of the bacterial cells during growth. Therefore, it is believed that the attachment of the subunits to the cell and to each other is done by way of weak non-covalent chemical bonds (Sara & Sleytr, 2000).

Amino acid analyses of S-layers reveal that the proteins are highly conserved even among phylogenetically different microorganisms (Beveridge, 1994). Indeed, there is a high content of acidic and hydrophobic amino acids, in particular the amino acid lysine. Also, the isoelectric points of the majority of the amino acids making up S-layers are in the weakly acidic pH range.

A remarkable feature of many S-layers is their glycosylation (Sleytr & Beveridge, 1999). In Gram-positive bacteria and Archaea, the sugars are covalently bound to the amino acid residues involved. Such sugars often form chains typically referred to as glycan chains. The glycan chains, when present, are 20 to 50 identical repeating carbohydrate units, similar in a way, to the glycan chains of the LPS of Gram-negative bacteria. In addition to these glycan chains, S-layer proteins show core oligosaccharide chains covalently bound to the protein moiety by O-glycosidic bonds.

The rate of synthesis of S-layer subunits is strictly controlled (Sara & Sleytr, 2000). The subunits must be produced quickly and translocated rapidly across the cell wall and cell membrane in order to keep the bacterial cell completely surrounded by S-layers. To maintain such a rate of synthesis requires strong promoters for the S-layer gene to be transcribed. Indeed, the promoter of the S-layer gene of *Lactobacillus acidophilus* has been found to be twice as efficient as that of lactate dehydrogenase. This observation is particularly noteworthy given that the lactate dehydrogenase gene is considered by many to have a strong promoter to begin with. Most S-layer proteins are synthesized with an N-terminal secretion signal peptide allowing for the translocation across the cell membrane to occur. The signal peptide is removed from the S-layer proteins after translocation.

The functions and the importance of S-layers in pathogenicity are not completely understood. However, given that they are ubiquitous, rich in proteins, and metabolically expensive, it is assumed that S-layers provide the bacterium an advantage or adaptation in its specific environment. They are considered to be important in processes such as molecular sieving, adhesion, antigenicity, and phage reception.

An important trait of many bacteria producing S-layers is their ability to change S-layer proteins, leading to a modified cell surface chemistry. For example, *Campylobacter fetus* produces two types of S-layer protein (Dworkin & Blaser, 1997). This type of chemical modification at the cell surface translates into a significant antigenic variation which helps the pathogen avoid the host's immune response such as phagocytosis.

Several studies have been conducted on the importance of S-layers as virulence factors. The molecular sieving function of S-layers has been shown to be particularly important for pathogenic bacteria because accurate sieving may prevent host defense substances such as lytic enzymes, complement, antibodies, and various biocides from entering the bacterial cell (Sleytr & Beveridge, 1999). In the fish pathogen, *Aeromonas salmonicida*, the S-layer allows the bacterium to resist the antimicrobial activity of complement (Trust et al., 1993). The S-layers of the human pathogen involved in periodontal disease, *Campylobacter rectus*, protect the microbe against phagocytosis by inhibiting adherence to fibroblast cells of the gum (Borinski & Holt, 1990). In some cases, S-layers may actually promote adherence. For example, the S-layers of the intestinal pathogen, *Clostridium difficile*, is believed to help the bacterium adhere to the intestinal epithelium (Takeoka et al., 1991). Overall, S-layers seem to be significant virulence factors.

References

Abeyta, M., Hardy, G., and Yother, J. (2003). Genetic alteration of capsule type but not PspA type affects accessibility of surface-bound complement and surface antigens of Streptococcus pneumoniae. *Infect. Immun.* 71: 218-225.

Allen, P.M., Roberts, I., Boulnois, G.J., Saunders, J.R., and Hart, C.A. (1987). Contribution of capsular polysaccharide and surface properties to virulence of Escherichia coli K1. *Infect. Immun.* 55: 2662-2668.

Beveridge, T.J. (1994) Bacterial S-layers. *Curr. Opin. Struct. Biol.* 4:204-212.

Borinski, R. and S.C. Holt. 1990. Surface characteristics of *Wolinella recta* ATCC 33228 and human clinical isolates: correlation of structure with function. *Infect. Immun.* 58: 2770-2776.

Costerton J.W., Irvin, R.T., and Cheng, K.-J. (1981). The bacterial glycocalyx in nature and disease. *Annu. Rev. Microbiol.* 35: 299-324.

Costerton, J.W., Geesey, G.G., and Cheng, K.-J. (1978). How bacteria stick. *Sci Am.* 238: 86-95.

Costerton, J.W., Stewart, P.S., and Greenberg, E.P. (1999). Bacterial biofilms: a common cause of persistent infections. *Science* 284: 318-322.

Cross, A.S. (1990). The biologic significance of bacterial encapsulation. *Curr. Top. Microbiol. Immunol.* 150: 87-95.

Davey, M.E. and O'Toole, G.A. (2000). Microbial biofilms: from ecology to molecular genetics. *Microbiol. Molec. Biol. Rev.* 64: 847-867.

Dworkin, J. and M.J. Blaser (1997) Molecular mechanisms of Campylobacter fetus surface layer proteins expression. *Mol. Microbiol.* 26: 433-440.

Finlay, B.B. and Falkow, S. (1989). Common themes in microbial pathogenicity. *Microbiol. Rev.* 53: 210-230.

Gilbert, P., Das, J., and Foley I. (1997). Biofilms susceptibility to antimicrobials. *Adv. Dent. Res.* 11:160-167.

Moxon, E.R. and Kroll, J.S. (1990). The role of bacterial polysaccharide capsules as virulence factors. Curr. Top. *Microbiol. Immunol.* 150: 65-85.

Murray, P.R., Baron, E.J., Jorgensen, J.H., Landry, M.L., and Pfaller, M.A. (2007). *Manual of Clinical Microbiology*. 9th Edition. ASM Press, Washington, D.C.

Nelson, A.L., Roche, A.M., Gould, J.M., Chim, K., Ratner, A.J., and Weiser, J.N. (2007). Capsule enhances pneumococcal colonization by limiting mucus-mediated clearance. *Infec. Immun.* 75: 83-90.

Nicholson, A.M. and Glynn, A.A. (1975). Investigation of the effect of K antigen in *Escherichia coli* urinary tract infections by use of a mouse model. *Br. J. Exp. Pathol.* 56: 549-553.

O'Riordan, K. and Lee, J.C. (2004). Staphylococcus aureus capsular polysaccharides. *Clin. Microbiol. Rev.* 17: 218-234.

Parsek, M.R. and Singh, P.K. (2003). Bacterial biofilms: an emerging link to disease pathogenesis. *Annu. Rev. Microbiol.* 57: 677-701.

Pluschke, G. and Achtman, M. (1984). Degree of antibody-independent activation of the classical complement pathway by K1 *Escherichia coli* differs with O antigen type and correlates with virulence of meningitis in newborns. *Infect. Immun.* 43: 684-692.

Sara and Sleytr (2000) S-layer proteins. *J. Bacteriol.* 182. 859-868.

Sleytr, U.B. (1978) Regular arrays of macromolecules on bacterial cell walls: structure, chemistry, assembly, and function. *Int. Rev. Cytol.* 53: 1-64.

Sleytr, U.B. and T. J. Beveridge (1999) Bacterial S-layers. *Trends Microbiol.* 7: 253-260.

Takeoka, A., K. Takumi, T. Koga, and T. Kawata (1991) Purification and characterization of S-layer proteins from Clostridium difficile GAI 0714. *J. Gen. Microbiol.* 137:261-267.

Trust, T.J., M. Kostrzynska, and E. Levente (1993) High-affinity binding of the basement membrane protein collagen type IV to the crystalline virulence surface protein array of Aeromonas salmonicida. *Mol. Microbiol.* 7:593-600.

Wilkinson, J.F. (1958). The extracellular polysaccharides of bacteria. *Bacteriol. Rev.* 22: 46-73.

In: Molecular Aspects of Infectious Diseases
Editor: Mazen T. Saleh
ISBN: 978-1-61728-690-2

Chapter IV

An Introduction to Protein Secretion in Prokaryotes

***Johanna DeLongchamp*[1], *Garry Ferroni*[1,2], *and Mazen Saleh*[1,2]**
Department of Biology Laurentian University, Sudbury, Ontario, Canada, P3E 2C6[1]
Northen Ontario School of Medicine' Laurentian University, Sudbury, Ontario, Canada, P3E 2C6[2]

A bacterium's ability to successfully transport proteins and other molecules across its cellular envelope is extremely important to its survival. Cells must be able to take in nutrients, excrete waste and secrete structural elements and virulence factors all while maintaining a stable interior environment. In order to selectively allow the passage of certain molecules across their barriers, bacteria must express specific transporter systems to recognize and facilitate the translocation of specific substrates.

This chapter will review seven different protein secretion systems in prokaryotes. It is important to note that protein secretion differs between Gram negative and Gram positive bacteria due to differences in the composition of their barriers. Gram-positive bacteria have an inner plasma membrane followed by a thick peptidoglycan cell wall. Gram negative bacteria on the other hand are more complex; they also contain an inner plasma membrane followed by a peptidoglycan layer, but this layer is much thinner than that of Gram positive bacteria and is found within the periplasmic space separating the inner membrane from the outer membrane. The outer membrane acts as a molecular sieve to impede the access of harmful substances, such as antibiotics, to the plasma membrane while allowing small dissolved nutrients to pass easily. As such, the outer membrane adds an additional barrier for secreted proteins to cross before they may reach the extracellular milieu.

Type I secretion, found in both Gram positive and Gram negative bacteria, translocates proteins directly from the cytosol to the extracellular space via a one-step mechanism energized by ATP hydrolysis. Type II secretion is much more complex involving transport across the inner membrane by either the general secretory (Sec) pathway or the twin arginine translocation (Tat) pathway followed by secreton-dependent translocation across the outer membrane. The type III secretion system of Gram negative bacteria is a very complex apparatus used to directly infect host eukaryotic cells with effector proteins. Similarly, type

IV secretion also directly infects a host cell with effector proteins however the apparatus contains fewer subunits and has a less complex mechanism. The machinery of the type IV secretion system is simply a conjugation apparatus adapted for use in virulence. Type V secretion is a form of specific outer membrane transport and as such is only found in Gram negative bacteria. The three individual pathways that make up this system consist of polypeptides previously translocated to the periplasm (most likely via the Sec pathway) that contain all the information required for their translocation across the outer membrane without the use of already-existing translocators. The recently identified type VI pathway is found in pathogenic and non-pathogenic Gram negative bacteria, and may directly infect host cells in a manner similar to type IV. Lastly, the type VII secretion system is found in mycobacteria and is likely involved in pathogenesis, possibly aiding in mycobateria's escape from phagosomes.

All of these translocation systems transport different substrates, have different compositions and use different energy sources to drive the transport of folded and unfolded proteins outside of the cell.

Type I Secretion System (T1ss)

The Type I secretion pathway is the simplest of the secretion systems. Often referred to as a "channel-tunnel" system, the T1ss translocator is composed of three components which collectively form a one-step direct pathway from the cytosol to the extracellular space. While this secretion system is often associated with toxins and hemolytic enzymes such as the adenylate cyclase toxin (CyaA) of *Bordetella pertussis* and hemolysin (HlyA) of *Escherichia coli*, it is involved in the transport of a wide variety of unfolded proteins up to, and possibly greater than, 800kDa. In addition, type 1 secretion is also utilized to transport other compounds including sugars, bacteriocins and signaling molecules (Holland, Schmitt & Young, 2005). Interestingly, there is evidence that mammalian cells may secrete cytokines via a type 1-like system (Flieger *et al*, 2003). Secretion by this transporter is dependent upon ATP hydrolysis, as well as an uncleaved C-terminal secretion signal. While the signal sequence contains little to no conservation, proteins secreted by the type I pathway commonly contain telltale glycine rich repeats for calcium binding (Delepelaire, 2004). The C-terminal location of the signal peptide is telling of post-translational secretion and since proteins must be translocated in an unfolded state, chaperones are required for efficient secretion.

Architecture

In Gram negative bacteria the T1SS is composed of three subunits; the outer membrane factor (OMF), the membrane fusion protein (MFP) and an ATP binding cassette (ABC) protein. This system is sometimes called a half-transporter, as the ABC protein contains only one ATPase while most ABC transporters require two of these nucleotide binding domains. It is likely then, that the active form of this system is a dimer species (Holland, Schmitt & Young, 2005). The ABC protein is the only component to interact directly with the secreted protein and as such it is likely responsible for the specificity of secretion. As shown in Figure 4.1, the inner membrane ABC subunit associates with the cytoplasmic N-terminus of the

MFP. The MFP, approximately 140Å in length, is anchored within the inner membrane and spans the periplasmic space (Delepelaire, 2004). The OMF, which is enclosed by the MFP, forms a β-barrel pore within the outer membrane. The absence of any one of these components leads to the loss of type I translocation activity, suggesting a highly cooperative mechanism (Holland, Schmitt & Young, 2005). This observation is not surprising, as a system of only three components will likely rely heavily on each of its subunits.

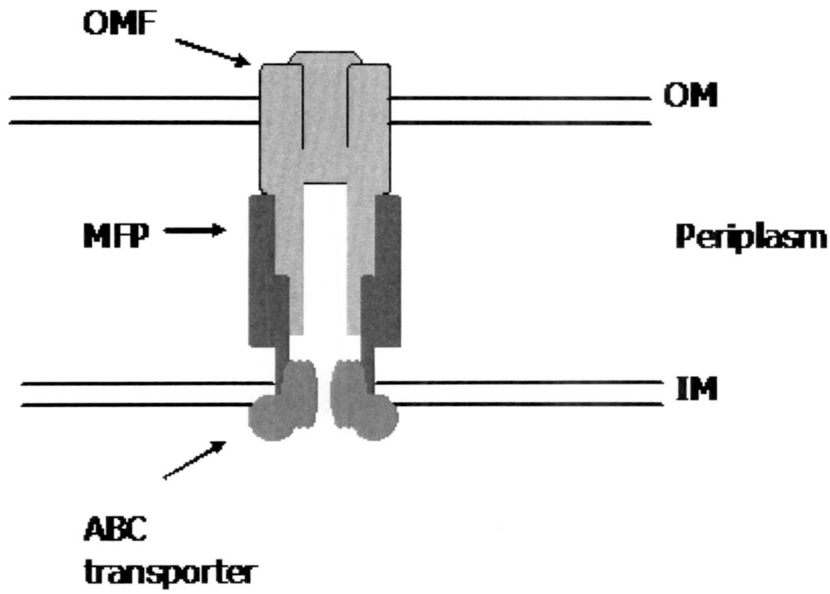

Figure 4.1 (a). A cutaway view of the three component translocator of type I secretion in Gram-negative bacteria. The trimeric outer membrane factor (OMF) spans the outer membrane (OM) and most of the periplasm. The multimeric membrane fusion protein (MFP)/adaptor protein surrounds the narrowed periplasmic end of the OMF as well as the integral inner membrane (IM) ABC protein.

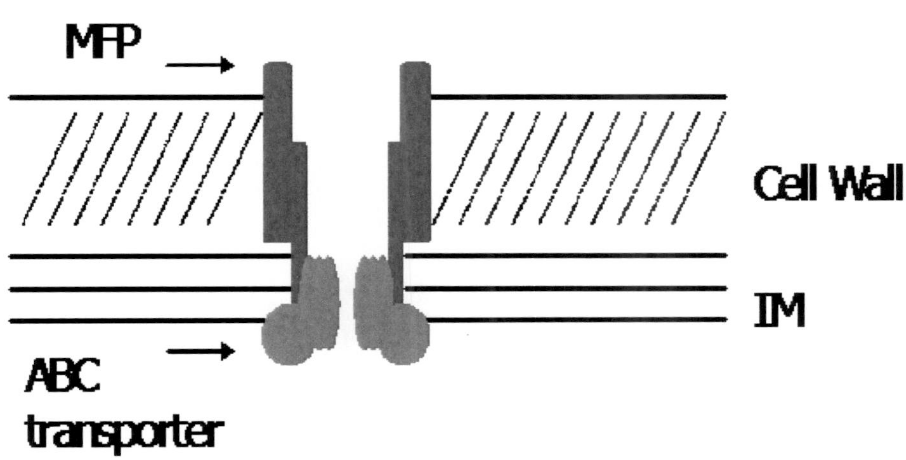

Figure 4.1 (b). A cutaway view of the type I translocator in Gram positive bacteria. Note that the MFP crosses the cell wall and is open to the extracellular milieu, there is no OMF.

A well-studied model of the T1SS is the hemolysin (Hly) system in *E.coli.* In this system the trimeric TolC is representative of the OMF, HlyD represents the MFP and HlyB is the ABC domain. The crystal structure of TolC reveals three protomers, each contributing four β-strands to form a cylindrical 12-strand β-barrel channel embedded in the outer membrane. This β-channel is approximately 40Å long and continues into a 100Å α-helical tunnel which projects into the periplasmic space. Unlike typical outer membrane β-barrel porins, TolC lacks an inward facing loop or plug to obstruct the passageway. The pore contains an average accessible diameter of 19.8Å and tapers at the periplasmic opening to a resting closed state of 3.9Å. Transition to an open state is likely to be induced by a conformational change via an iris-like rearrangement of the α-helices, resulting in a periplasmic opening of 16-20Å (Koronakis, Eswaran & Hughes, 2004).

Mechanism of Transport

Once the protein has been fully synthesized, chaperones within the cytoplasm bind to mask aggregation prone sites and prevent premature folding as well as proteolytic degradation. The chaperone-preprotein complex is recruited to the ABC domain where the secretion signal directly interacts with the transporter. Since the primary sequence of the transmembrane domains (TMDs) of ABC-transporters are less conserved than that of the nucleotide binding domains (NBDs), they are suggested to confer the specificity of transport. As such, it is likely that the secretion signal specifically interacts with the TMD (Delepelaire, 2004). Under normal resting conditions, the Walker A (P-loop) forms a 3_{10} helix rather than a classical helix, resulting in intramolecular bonding between a Walker A lysine and a Walker B glutamate residue, obscuring the ATP binding site. The docking of a protein's signal peptide causes a conformational change within the Walker A site, disrupting the intramolecular bond and shifting the structural organization to that of a classical helix. With the ATP binding site no longer obstructed, ATP is recruited to the ATP binding cassette (Holland, Schmitt & Young, 2005). It has been speculated that ATP binding may trigger homo-dimerization of the ABC protein. It is important to note here that without the presence of a protein to be transported, ATP cannot bind to the NBD and so the transporter avoids unnecessary ATP binding. The binding of the chaperoned complex also triggers the rearrangement of adaptor protein HlyD, which surrounds and consequently opens the periplasmic entrance of TolC, creating a continuous tunnel to the extracellular milieu. Simultaneously, upon binding ATP, the ABC protein releases the previously bound protein into the translocation pathway and proton motive force (PMF) is responsible for transporting protein the remainder of the way. Once transport is complete, it is hypothesized that ATP is hydrolyzed to reset the transporter to a de-dimerized 3_{10} helical state once again. This sequence of events is unusual for an ABC-transporter, as ATP is typically hydrolyzed to supply the energy for transport (Holland, Schmitt & Young, 2005).

Protein Folding

It is unclear at what point the transported pre-protein folds into a functional protein. The TolC subunit has been suggested to be a "folding cage" in which the protein has a safe enclosure to properly fold (Holland, Schmitt & Young, 2005). Mutagenesis studies have shown that TolC mutants secrete HylA in less active and aggregate-prone forms, as well as at slower secretion rates (Delepelaire, 2004). It has also been proposed that unfolded protein folds on the bacterial extracellular surface. Clearly there are no chaperones to aid in folding outside of the cell. However, as mentioned earlier, several type I polypeptides contain C-terminal glycine rich repeats which bind calcium ions. As the lipopolysaccharide (LPS) rich bacterial surface likely contains an abundance of Ca^{2+}, these repeats may serve an autocatalytic role in protein folding. Studies have shown that heterologous proteins expressing the glycine-rich C-terminus of HlyA fold correctly as well (Holland, Schmitt & Young, 2005).

Summary

- The type I secretion pathway is simple channel-tunnel system which extends from the cytoplasm to the extracellular milieu and is involved in the transportation of several proteins and other molecules.
- There are only three components to this system and the deletion of any one component abolishes transport ability.
- This type of protein secretion is reliant upon ATP and PMF energy to drive the translocation.
- The straightforward three component system also depends on important conformational changes, and most likely dimerization, to function.
- The proteins are transported post-translationally in an unfolded state and are probably guided to the translocation machinery by a chaperone.
- The C-terminal secretion signal interacts directly with the translocation machinery, specifically with the TMD of the ABC subunit, resulting in a specificity of secretion.
- It is currently unclear whether the substrate folds within the channel or once outside the cell.
- The Ca^{2+} binding glycine repeats at the C-terminus likely play an important part in protein folding.

Type II Secretion System (T2ss)

The type II system is a two-step mechanism of protein secretion in Gram negative bacteria. The protein is first transported across the cytoplasmic inner membrane into the periplasm by either the General Secretory Pathway (Sec) or the Twin Arginine Translocation (Tat) pathway. Transport across the outer membrane is then achieved by one of several terminal branches. The Sec and Tat pathways are also found in Gram positive bacteria, however with no outer membrane the second step of the type II mechanism is not required. The T2SS exports a large variety of proteins including lipases, chitinases, acyltransferases, amylases, cellulases, proteases, phospholipases, acid and alkaline phosphatases, and nucleases among several others (Cianciotto, 2004). Four lines of evidence implicate the type II pathway in the pathogenesis of several bacteria. To begin, the genes for a functional T2SS are found in a variety of notable pathogens. Secondly, the enzymes secreted by this system, as listed above, are by and large degradative in nature suggesting that they may cause significant damage to target systems (mammalian, plant etc). In addition, several individually identified proteins known to be transported by this pathway are recognized virulence factors (cholera toxin of *Vibrio cholera*, exotoxin A of *Pseudomonas aeruginosa*). Lastly, directed mutagenesis of the type II machinery assuages virulence (Cianciotto, 2004). Furthermore, it should be noted that this system helps promote bacterial growth in intracellular niches and also secretes effectors that contribute to the subversion of the innate immune system (Cianciotto, 2004).

General Secretory Pathway (Gsp)

As suggested by the name, the Sec translocation pathway is a widespread, evolutionarily conserved transport system. In fact the Sec translocon can be found in both prokaryotic and eukaryotic systems, however only the bacterial complex will be discussed here.

Architecture

The Sec pathway is composed of several components and for clarity as well as consistency, the well studied Sec pathway of *Escherichia coli* will be used as a model (Figure 4.2a). SecB is a cytosolic chaperone which recognizes and binds to proteins destined for export via the Sec pathway. The chaperone is responsible for delaying folding, preventing aggregation, and targeting the protein to the Sec translocase located within the inner membrane. The Sec translocase, which can only transport proteins in an unfolded state, is composed of the heterotrimeric SecYEG, which forms the actual channel for translocation, and a SecA ATPase to drive the translocation. Several accessory proteins have also been characterized including SecDF, YajC, YidC and a signal peptidase (Stephenson, 2005).

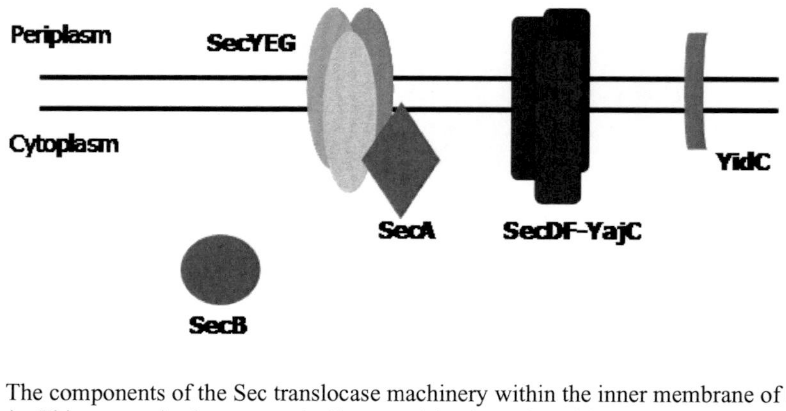

Figure 4.2 (a). The components of the Sec translocase machinery within the inner membrane of Gram-negative bacteria. This system is also present in Gram positive bacteria, with the transmembrane components simply spanning both the cytoplasmic membrane and the cell wall. The unfolded pre-protein transverses the membrane by passing through the channel formed by the SecYEG heterotrimer while SecA hydrolyzes ATP to feed the polypeptide though the channel.

SecB is a 16.6kDa homotetramer, composed of a dimer of dimers, with two 70Å binding grooves on opposite sides (Kim & Kendall, 2000). It is proposed that upon binding of SecB to the pre-protein, the substrate wraps itself around the chaperone to occupy both grooves. SecB substrates do not share sequence homology, however a short binding motif composed of nine aromatic and charged residues is partly responsible for the specificity of binding. In addition, charged residues around the binding groove could act as a selectivity filter for substrate binding. Indeed, it has been shown that SecB associates transiently with proteins not destined for the Sec translocase, but forms a tight complex with those that are (Dekker *et al*, 2003). It should be noted, however, that not all bacterial genomes contain a homologue for SecB and that some proteins can indeed efficiently target the translocase on their own (Stephenson, 2005).

SecA, a 100 kDa dimeric ATPase, is the receptor for secretory proteins of the translocase and exists in membrane-bound and cytosolic forms. Without SecA, all protein secretion via the general secretory pathway is abolished (Kim &Kendall, 2000). SecA interacts with SecYEG, acidic phospholipids, SecB and with both the signal peptide and mature regions of the pre-protein. The N-terminal ATPase domain of SecA contains the protein binding site as well as two proposed nucleotide binding domains (NBD1 and NBD2). NBD1 alone plays a catalytic role while NBD2 acts strictly as a regulator. The C-terminus of SecA contains a Zn^{2+} coordinated binding site for SecB as well as the lipid binding sites (Mori & Ito, 2001). The ATPase activity of SecA on its own is in fact very low and requires direct association with a lipid environment, membrane-bound proteins (SecYEG) and a protein precursor to stimulate full activity (Kim & Kendall, 2000).

The heterotrimeric SecYEG forms an hour glass shaped channel spanning the inner membrane. The cytosolic face of the channel has a diameter of 20-25Å and proceeds to narrow to 5-8Å at the constriction site before widening again at the external face. The walls of the channel are lined with uncharged amino acids which likely function to minimize substrate interactions with SecYEG. The constriction site however is lined with hydrophobic residues which have been proposed to form a seal around the substrate during transport, thus maintaining membrane integrity. A small loop from the α-subunit also projects across the constriction site, acting as a plug which is likely displaced upon association of the secretory protein with the translocase (Stephenson, 2005).

The membrane-bound SecDF-YajC accessory complex plays a role in membrane cycling of SecA, translocon assembly and the clearing of misfolded proteins and cleaved signal peptides (Stephenson, 2005). There is also evidence that SecD and SecF aid in the prevention of reverse translocation of the substrate (Stathopoulos *et al*, 2000) The integral membrane accessory protein YidC is involved in substrate membrane insertion, although YidC depletion only has a minimal effect on the insertion of Sec-dependent membrane proteins (Stephenson, 2005). Finally, the membrane bound signal peptidase is responsible for the cleavage of the signal sequence releasing the mature protein for insertion into inner membrane or into the periplasmic space (Kim & Kendall, 2000).

Mechanism of Transport

Secretion is directed by an N-terminal signal peptide composed of a core of ten or more hydrophobic residues flanked by a positively charged polar N-domain and a hydrophilic polar

C-domain (Stephenson, 2005). As described earlier, cytosolic chaperones such as SecB bind to the pre-protein to prevent folding and aggregation. SecB is also responsible for directing the nascent polypeptide to the Sec translocon located within the inner membrane. As the secretion signal is located at the N-terminus, transport of the protein can either be co-translational or post-translational (Stephenson, 2005). SecB targets the bound protein precursor to the inner membrane translocon where SecA, in association with SecYEG, binds the pre-protein causing SecB to release its substrate. ATP binding by SecA then results in the insertion of the first 20-30 N-terminal amino acids of the pre-protein, along with carboxy terminal of SecA itself, into the translocation channel of SecYEG. ATP hydrolysis via NBD1 of SecA then leads to the release of pre-protein and the deinsertion of SecA, allowing the ATPase to bind another segment of the precursor and another ATP molecule for the insertion of further 20-30 amino acids. This cycling of SecA in the "hand-over-hand" translocation of the precursor continues until a sufficient proportion has entered the pore, at which point the remainder of the polypeptide is translocated using proton motive force (PMF) (Stephenson, 2005). While both ATP hydrolysis and PMF are used to drive translocation in this pathway, only ATP is absolutely essential, with PMF only affecting the rates of translocation. Once the precursor has transversed the inner membrane it remains bound to the inner membrane via the H-domain of the N-terminal signal peptide. The most C-terminal segment of the signal peptide, which contains a specific cleavage site, is then cleaved by one of three integral membrane signal peptidases (Stathopoulos *et al*, 2000). Type I signal peptidase (SPase I), belonging to the serine protease family, releases proteins from the inner membrane so that they may reach their periplasmic or extracellular destinations (Tuteja, 2005). SPase II cleaves lipoprotein precursors and the invariant cysteine residue of the cleavage sequence is modified to become the retention signal so that the mature protein may insert itself in the membrane (Tjalsma *et al*, 1999). SPase IV is a specific prepilin peptidase for type IV pili (Stathopoulos *et al*, 2000).

Protein Folding

Once the signal peptide is cleaved, the mature domain is released for folding into the native conformation, often aided by folding catalysts found in the periplasmic space. For example, the Dsb system of *E. coli* is responsible for stabilizing proteins by forming disulfide bridges, while the peptidyl-prolyl isomerases (PPIases) aid in the isomerization of proline residues needed for correct folding (Stathopoulos et al, 2000). In addition to these folding catalysts, chaperones are also present to bind exposed hydrophobic sequences thus preventing aggregation and proteolysis, as well as to act as shuttles to the outer membrane transport machinery. However, none of the classical chaperones of Hsp60 and Hsp70 families are found in the periplasm. These chaperones are reliant upon ATP activity and as there is no available ATP within the periplasmic space, protein folding must be supported by another group of chaperones (Missiakas & Raina, 1997). Indeed three proteins with general chaperone activity have been identified in the periplasm; Skp, DegP and SurA (Sklar *et al*, 2007). However, depending on the properties of specific proteins and their mode of transport across the outer membrane, some proteins may need to remain partially unfolded for transport across the outer membrane.

Twin Arginine Translocation (Tat) Pathway

The Tat pathway is not as widespread as the general secretory pathway but instead transports a separate and specific class of proteins. Unlike the Sec pathway, the Tat pathway translocates fully folded proteins up to, and possibly exceeding, 100 kDa across the cytoplasmic membrane. The ability to accommodate large folded substrates allows the Tat translocase to transport proteins containing cofactors such as iron-sulfur clusters, which would otherwise not be available for incorporation if the protein were to fold in the periplasm. In addition, proteins exhibiting fast folding kinetics are transported via the Tat translocase rather than the cell expending energy trying to prevent folding with a chaperone (Muller, 2004). Again, while the Tat pathway is found in prokaryotes and the thylakoid membranes of plants, only the bacterial system is discussed here.

Architecture

The size of the Tat translocation pore must be highly adaptable to allow transport of monomeric and oligomeric proteins ranging from 10 to 100 kDa, while still maintaining membrane integrity. The Tat translocon is not known in as great detail as the Sec machinery, however it is known that there are three subunits; TatA, TatB, and TatC (Figure 4.2b) expressed in a relative ratio of approximately 20-30:1:0.4, respectively, all of which are encoded by the *TatABC* operon. Deletion studies have also showed that the removal of any one of these components leads to diminished translocon function. With several copies of TatA being expressed and the tendency of this substrate to homo-oligomerize, it seems that TatA forms the adjustable pore for translocation across the cytoplasmic membrane (Muller, 2004). A *TatD* gene has also been identified as expressing a cytoplasmic protein with no effect on Tat transport, in addition to a *TatE* gene which encodes a TatA homologue (Robinson & Bolhuis, 2001).

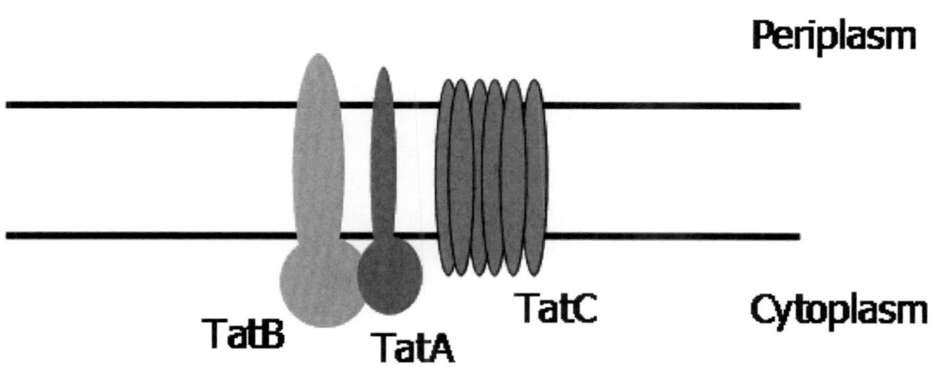

Figure 4.2 (b). The key components of the Tat translocase within the inner membrane of Gram negative bacteria, as encoded by the TatABC operon. Once again, this system is found in Gram-positive bacteria with the components spanning the cell wall in addition to the cytoplasmic membrane. TatC likely interacts with substrate first, resulting in activation and co-operative assembly of the system. TatC and TatB associate while TatA homo-oligomerizes to form a flexible pore, specific for each substrate. Transport across the membrane is driven by PMF.

Both TatA and TatB are anchored in the cytoplasmic membrane by an N-terminal single pass transmembrane helix. Each contains a glycine residue just after the transmembrane segment which confers flexibility to these subunits. TatA and TatB have been repeatedly isolated in complex together as a 600 kDa unit and fusion experiments of these two subunits had no effect on their function. TatC, in contrast, is more complex and is predicted to span the membrane six times. Interestingly, TatB-TatC fusions are also shown to be entirely functional (Robinson & Bolhuis, 2001). These observations suggest that TatB associates with both TatA and TatC in the functional translocon. While it seems likely that the translocation machinery in its entirety is composed solely of the TatABC complex, we cannot know for sure until the complete and functional translocase has been purified.

Mechanism of Transport

The approximate 37-residue signal peptide of Tat substrates contains twin arginine residues, thus its namesake. The R-R-x-F-L-K motif, where x is a polar residue, directs the substrate to the Tat translocase. Double mutation of the two arginine residues abolishes transport while mutation of only one results in a slower rate of translocation. Interestingly, the presence of the R-R motif does not necessarily correlate to transport via the Tat pathway, as several predicted substrates in *Bacillus subtilis* containing this motif did not actually target the Tat pathway (Muller, 2004). Another interesting aspect of the Tat signal peptide is the "Sec avoidance signal" it contains. This avoidance signal is a basic residue located in the C-terminus of the signal peptide which helps the substrate avoid the ubiquitous Sec pathway and reach its appropriate transporter (De Buck, Lammertyn & Anne, 2008).

While the exact mechanism of Tat transport has yet to be unveiled, several similar pathways have been proposed. Alami *et al* suggest that the substrate is targeted first to TatC which results in the association of TatB-TatC and the subsequent transfer, dependent on PMF and catalyzed by TatB, of the substrate from TatC to the TatA pore. Thus TatB plays an intermediate role between substrate recognition and subsequent translocation. TatA will homo-oligomerize and form a pore only in the presence of a H^+ gradient and when a substrate has been recognized and bound by the TatBC complex (Alami et al, 2003; Muller, 2004). In addition, it has been recently discovered that several Tat substrates have chaperones to prevent their interaction with the translocase machinery until they have successfully folded. For example, accessory protein TorD binds, and thus obscures, the signal peptide and part of the mature sequence of cofactor-less and partially folded tri-methylamine N-oxide (TMAO) preventing it from targeting the Tat translocase and maintaining a cofactor-competent state. Insertion of TMAO's cofactor leads to the release of TorD, allowing the now fully folded and cofactor-containing enzyme to target the Tat pathway (Robinson & Bolhuis, 2001).

In summary, upon translation of a Tat substrate the signal peptide may or may not be obscured by a chaperone for quality control of folding and cofactor insertion before it targets the Tat translocon. The R-R signal peptide averts the Sec machinery and interacts with membrane bound TatC, resulting in the recruitment and association of TatB. In response to TatC-TatB association with the substrate, and in the presence of a proton gradient, TatA homo-oligomerizes into a transmembrane pore of appropriate size. TatB then, using the transmembrane proton motive force (PMF), catalyzes the transfer of the Tat substrate to the TatA translocation pore. The substrate is consequently translocated, energized by PMF, across the cytoplasmic membrane and either cleaved by a signal peptidase for release or remains uncleaved and inserts into the membrane (Berks, Palmer & Sargent, 2005).

Outer Membrane Transport

Whether secreted by the Sec or Tat pathway, all proteins within the periplasmic space are folded or partially folded which, in addition to the lack of ATP, proposes a challenge for their specific transport across the outer membrane. All transport mechanisms that bring Sec and Tat substrates across the outer membrane, as the final step in secretion, are called terminal branches. The autotransporter (AT), two-partner secretion (TPS) and chaperone/usher (CU) pathways likely utilize the energy from freshly secreted partially folded and unfolded polypeptides to drive transport and thus do not rely on energy harnessed from the inner membrane (Thanassi *et al*, 2005). The main terminal branch of outer membrane transport is known as the type II secretion system or the secreton-dependent pathway. The type II apparatus consists of twelve core components, as shown in Figure 4.2c, including an outer membrane secretin, as well as an inner membrane platform composed of a cytoplasmic ATPase, major and minor pseudopilins, and stabilizing transmembrane proteins (Cianciotti, 2005). The outer membrane pore is suggested to be formed by 12-15 secretin subunits. Indeed biochemical and electron microscopy have shown that multimeric secretin proteins form a ring structure in the outer membrane of *Pseudomonas aeruginosa* with a central cavity diameter of 95Å (Filloux, 2004). With the pore size so large, a mechanism for opening and closing is essential to avoid cell death. A system homologous to the TonB system has been suggested as a gating mechanism for the outer membrane pore. TonB is an inner membrane-bound protein with a large periplasmic domain capable of interacting with and opening its corresponding outer membrane porin via a conformational change. It has also been suggested that the pseudopilus could be responsible for either acting as a plug in blocking the pore, or actively pushing the exoproteins through the pore while a different gating mechanism altogether controlled entry to the outer membrane pore (see Figure 4.2c) (Filloux, 2004).

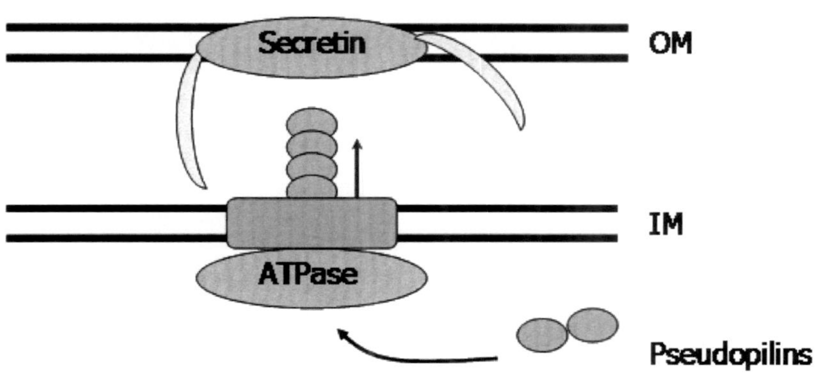

Figure 4.2 (c). The simplified structure of the type II apparatus / main terminal branch of outer membrane transport. Pseudopilus assembly is achieved through the cyclic inner membrane ATPase and may serve the purpose of corking the inactive secreton or pushing secretory proteins through the cyclic secretin outer membrane pore. Note the proposed gating mechanism which is depicted in the closed and open states (left and right respectively). OM transport relies on PMF directly as well as on ATP energy harnessed from the inner membrane.

Summary

- The type II secretion pathway is a two step mechanism of protein secretion in Gram negative bacteria.
- Transport across the inner membrane (IM) is accomplished by either the Sec or Tat pathway in unfolded or folded states respectively.
- The Sec pathway relies on ATP and PMF energy while the Tat pathway relies only on PMF energy.
- Both pathways rely on an N-terminal secretion signal and both pathways transport soluble and membrane proteins.
- The Tat pathway must be adaptable to adjust to the variety of proteins exported this way.
- Outer membrane (OM) transport is accomplished by terminal branches of the T2SS, with the main terminal branch being the secreton.
- As with the Tat pathway, the secreton must be adaptable for the diverse assortment of proteins to be translocated.
- As there is no ATP available in the periplasm, OM transport relies on PMF and energy harnessed from the IM.
- The exact mechanism of OM transport is still unresolved.

Type III Secretion System (T3ss)

The type III secretion system, also called the injectisome, is found in Gram negative bacteria that interact with other organisms, either as symbionts or more commonly as pathogens. The injectisome is a long hollow needle which, when bacteria dock onto a eukaryotic host cell, directly inject anywhere from six to more than twenty different bacterial effector proteins into the cytoplasm of the host cell (Cornelis, 2006). Thus this complex system skillfully couples protein secretion with translocation across eukaryotic host cell membranes and the effector protein is never exposed to the extracellular environment. These translocated effectors modulate the biochemical activities of the host to serve the main purpose of ensuring the survival and transmission of the colonizing bacteria. More specifically, of the more than 100 effectors characterized to date, they have been found to inhibit phagocytes, induce apoptosis, down-regulate the pro-inflammatory response, aid in host cell invasion, inhibit autophagy and modulate intracellular traffic (Cornelis, 2006). Generally speaking, it is not beneficial for bacteria to fatally harm their host as it endangers their own survival. However sometimes, as in the case of an already weakened immune response, the balance of the microbe-host interaction is disturbed and T3SS-pathogens can cause a broad spectrum of disease (Galan & Collmer, 1999). Infection by pathogens such as enteropathogenic and enterohemorrhagic *E. coli*, *Shigella*, *Salmonella*, *Yersinia*, *B. pertussis*, *P. aeruginosa*, and *Chlamydia* species cause intestinal disease, plague, enteric fever, pneumonia, whooping cough, urinary tract/wound infections, septicemia, endocarditis and a variety of other ailments (Coburn, Sekirov & Finlay, 2007). Aside from the expansive pathogenesis and complexity of the T3SS, there are three distinguishing characteristics of this system; the absence of a cleavable signal peptide, the absolute requirement for chaperones and the necessity of host cell contact (Galan & Collmer, 1999).

Architecture

The injectisome is structurally similar to the bacterial flagellum and the two likely share a common ancestor. In fact, the flagellum and injectisome share analogous basal body composition with the flagellum being topped by a hook for motility while the injectisome is topped with a needle instead (Journet, Hughes & Cornelis, 2005). The entire injectisome complex is composed of approximately 25 proteins, making it the most complex of all known secretion systems (Galan & Collmer, 1999). The base of the system is composed of two pairs of rings that span the inner and outer membranes, joined together by a rod (Figure 4.3). The needle structure, which varies in length from 45 to 80 nm, associates with the outer membrane ring and protrudes from the bacterial surface. In some bacteria the needle may be extended in length by the addition of filaments which might also aid in host cell attachment (Journet, Hughes & Cornelis, 2005; Coburn, Sekirov & Finlay, 2007). The external diameter of the needle is 8-13nm with the internal diameter being between 2-3nm, clearly too small to pass folded proteins. Interestingly, not only does the secretion system translocate effector proteins but it also translocates regulator and translocator proteins as well as its own structural constituents (Journet, Hughes & Cornelis, 2005). The Sec system previously discussed is responsible for assembling the base and export apparatus.

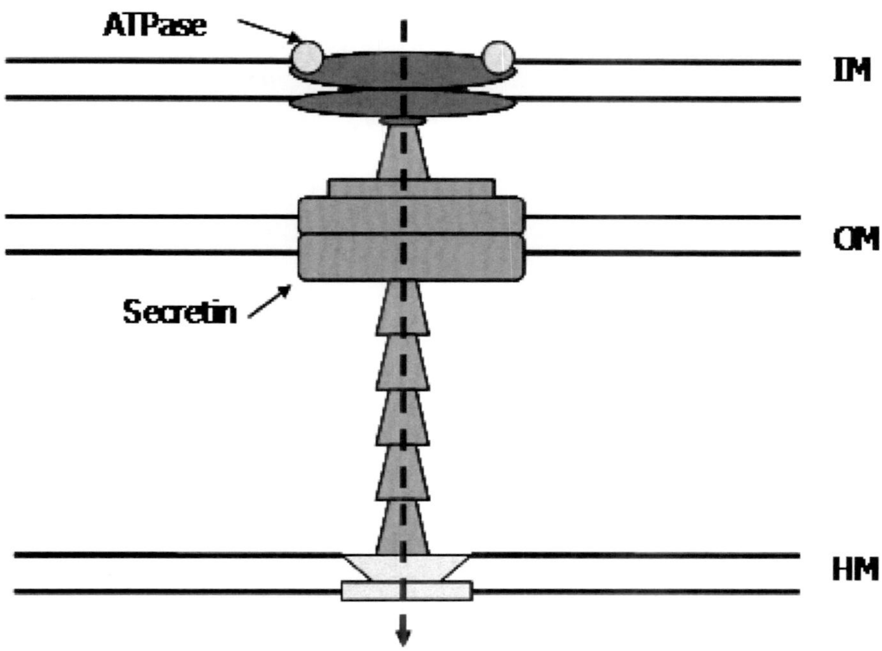

Figure 4.3. The needle complex of the T3SS that translocates effector proteins directly into the host cell. The secretin complex within the outer membrane secretes structural components to build the needle. Once the cell makes contact with the host organism, the system secretes translocator proteins to create a pore in the host membrane and allow the direct injection of effector proteins into the host cell's cytoplasm.

The distal needle however, is assembled by the sequential secretion of the individual subunits via the T3SS followed by the assembly of the subunits into a cylindrical helix with a central channel. It is currently unknown if the needle grows from the tip as with flagella or from the base as with frimbriae, however the model of tip-growth is preferred. Another uncertainty of the T3SS is the energy source with which translocation is powered. Previous studies have shown that disruption of the proton gradient interferes with the assembly of injectisome subunits (Galan, 2008). In addition, an inner membrane bound protein of the translocator contains sequence similarity to the bacterial F0F1 proton-translocation ATPase (Galan & Collmer, 1999). Based on these and other finding it can be suggested that protein secretion via the type III mechanism involves both ATP hydrolysis and PMF energy as driving forces.

Mechanism of Transport

Translocation is directed by a loosely defined N-terminal signal peptide with little to no sequence homology. In fact, substrate recognition is likely based on a disordered peptide structure with RNA signals possibly playing a secondary role (Journet, Hughes & Cornelis, 2005). Most, if not all, secretions of the type III system are post-translational and with the channel being so narrow chaperones are generally required to maintain proteins in an unfolded state. More specifically, there are three classes of chaperone; class I which assist effectors, class II which assist translocators and regulators, and class III which assist structural elements (Cornelis, 2006). Beginning with class I; these chaperones recognize a chaperone binding domain (CBD) approximately 100 amino acids downstream of the signal peptide. This interaction is specific as noted by the fact that the genes for both the chaperone and its substrate are often contained within the same operon. The small dimeric chaperones are acidic and play important anti-toxicity as well as targeting roles. In the absence of a chaperone, it's substrate will not be transported either due to proteolysis and aggregation susceptibility or loss of targeting ability, while the substrates of other chaperones are not affected (Journet, Hughes & Cornelis, 2005). The class II chaperones are strictly associated with translocator proteins which are hydrophobic and thus toxic within the cytosol. The main purpose of class II chaperones is the neutralization of these toxic substrates. Finally class III chaperones function in protecting structural elements from premature polymerization and self-association within the cytosol. These chaperones may also guide structural subunits to the export machinery in addition to their protective role (Cornelis, 2006). None of these chaperones are exported through the T3SS and are likely recycled. Research, in fact, shows that chaperones have to ability to behave as regulatory proteins once they release their cognate substrates. The recently liberated chaperones may play a significant role in coupling gene regulation to the ordered translocation of secreted proteins, acting as molecular clocks timing secretion to specific stages in translocator assembly and the secretion process in general (Journet, Hughes & Cornelis, 2005).

Upon the attachment of the T3SS-containing bacterium to the host eukaryotic cell, specific proteins at the bacterial surface, likely adhesins, sense the interaction, act to bridge the two cells and promote the passage of translocator proteins which subsequently result in the formation of a pore in the host membrane. With a direct route now open to the host cell, effector proteins with the purpose of disrupting the host cytoskeleton are translocated (Cornelis, 2006). These effectors manipulate Rho GTPases, which are important in cytoskeletal functioning, as well as directly interfering with globular and filamentous actin (Coburn, Sekirov & Finlay, 2007). With a compromised cytoskeleton, the host cells are nearly effortless to invade and colonize. Once the bacteria make themselves at home, they secrete additional effectors to manipulate their host further and ensure their survival and proliferation. The activation and functioning of the already assembled injectisome occurs rapidly, with half of the effector proteins being secreted within 240 seconds of the initial host contact. This hurried response is essential to avoid the host cell's phagocytic activity as phagocytes generally engulf invading bacteria within one minute (Cornelis, 2006). The inflammatory response can also be elicited or inhibited through intracellular signaling interference by the bacteria (Coburn, Sekirov & Finlay, 2007). Invariably, damage to host tissues results from cytotoxicity, apoptosis, necrosis and the breakdown of tissue barriers.

The currently proposed model for driving energy in this system involves both ATP and PMF. It is suggested that ATP hydrolysis is involved in the removal of chaperones, the unfolding of secretory proteins and the presentation of these proteins to the translocation machinery while PMF in responsible for guiding the unfolded proteins through the translocation pathway into the host cell (Galan, 2008). It has also been noted that the energy "stored" in unfolded proteins may also aid translocation through the needle.

Summary

- The type III secretion system is a very complex pathway in which bacterial effector proteins are transported directly from the cytosol of the bacterium to the cytosol of eukaryotic host.
- This system is found primarily in pathogenic Gram negative bacteria and is involved in a wide variety of disease and infection.
- The architecture of the T3SS bears significant similarity to that of the flagellum, suggesting a common ancestor.
- The T3SS secretes four types of proteins: structural proteins, regulator proteins, translocator proteins and effector proteins.
- With the translocation pathway within the tunnel being so small, all proteins must be translocated in an unfolded state.
- There are three different chaperone classes (I, II, III) responsible for the different secretory proteins.
- Transport is dependent upon an uncleaved N-terminal signal peptide, chaperones and host cell contact.
- Some chaperones may play a dual role by acting as regulators for transcription once they have released their cognate substrates.
- Secreted effectors manipulate the host cell to allow further colonization, invasion and subversion of the host's defenses.
- Transport via the T3SS requires ATP and PMF energy to function.

Type IV Secretion System (T4ss)

The type IV system is broadly distributed among both Gram positive and Gram negative bacteria. This system is used to translocate DNA as well as proteins in a one-step mechanism across bacterial envelopes. In fact, the T4SS can be broken down into three functional subgroups: conjugative transport systems, effector translocator systems, and DNA uptake and release systems (Christie & Cascales, 2005). The conjugative system is used to transfer mobile DNA to a bacterial recipient while the DNA uptake and release system exchanges DNA with the environment. Both of these systems lead to gene acquisition and may enable the invading pathogen to cope with its changing environment. For example, these mobile DNA elements often confer antibiotic resistance. In addition, the conjugative pili may contribute to biofilm formation and thus promote colonization of human tissues. The effector translocator system transfers DNA and proteins to eukaryotic hosts and, as the more relevant system to this chapter, will be the focus from here on. The delivery of effector molecules into a eukaryotic host enables the invading bacteria to suppress defense mechanisms and to promote intracellular changes that will support the survival of the pathogen. More specifically, effector proteins are capable of promoting the synthesis of nutrients for the invading pathogen in addition to interfering with intracellular signaling molecules with the purpose of inducing morphological changes in the host cell. *Bordetella pertussis, Bartonella henselae, Helicobacter pylori*, and *Legionella pneumonphila* are examples of the numerous pathogens that use type IV effector translocators during infection (Christie & Cascales, 2003).

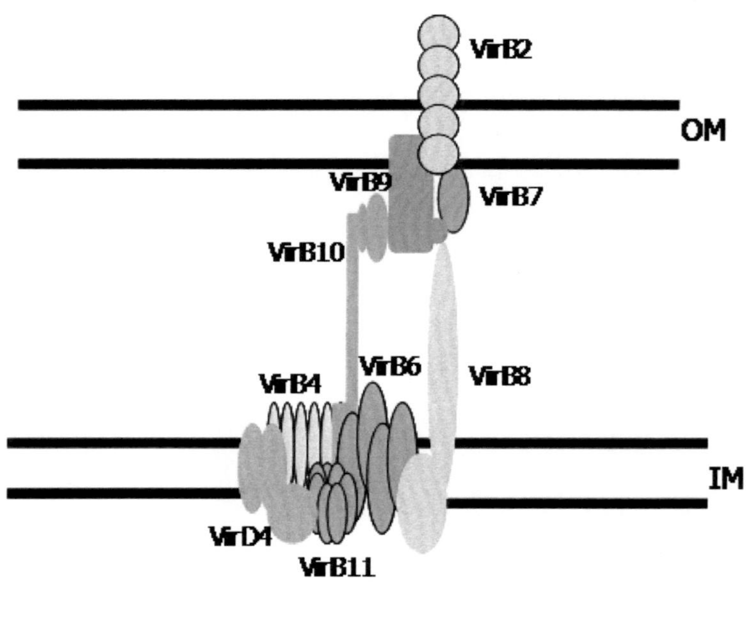

Figure 4.4. The type IV secretion apparatus of *A. tumefaciens*. The energy complex (VirD4, VirB4 and VirB11) is found at the inner membrane while VirB10, hypothesized to harness energy from the inner membrane for outer membrane secretion, spans the periplasmic space to interact with outer membrane bound subunits. It is likely that the VirB7 and VirB9 oligomerize to form a channel while VirB10 also oligomerizes to surround the channel and make up the outer wall. VirB1, VirB3 and VirB5 have been deleted for simplicity. (Adapted from Chrisitie & Cascales, 2005).

Architecture

The type IV system is fairly complex, composed of approximately twelve subunits. The transporter can be summarized into two major components; the pilus and the transmembrane complex. The polar-localized T4SS of *Agrobacterium tumefaciens* is a well studied conjugation apparatus adapted to deliver effectors and is consequently the model system for type IV secretion. *A. tumefaciens* translocates tumor-inducing DNA (T-DNA) along with other virulence factors to host plant cells, resulting in the formation of crown galls which is another term for plant tumors (Christie & Cascales, 2005). This system is formed by twelve proteins annotated VirB1 through VirB11 and VirD4, and is therefore referred to as the VirB/D4 system (Figure 4.4). Other T4SS-expressing bacteria may contain homologs for only a subset of the VirB/D4 components or may have their own additional components. This system contains an energy sub-complex composed of VirD4, VirB11 and VirB4. These three subunits, which have been shown to bind ATP, are individual ATPases involved in secretion and complex assembly (Christie & Cascales, 2005; Fronzes *et al*, 2009).VirB1, VirB3, VirB6, VirB7, VirB8, VirB9 and VirB10 form a complex which spans both the inner and outer membranes and may serve as the translocation pore for substrates (Backert, Fronzes & Waksman, 2008). Protein-protein interactions have confirmed a core complex consisting of VirB7, VirB8, VirB9, and VirB10 however solubilization and purification produced a three component core complex composed of VirB7-VirB9-VirB10 with a 1:1:1 stoichiometry. These components were each present as 14mers and structured into a ring-like, multilayered formation. The core complex is both 185 Å in height and 185 Å in diameter. The internal walls of the main body forms a chamber that is 110 Å at its widest point, constricting to 10 Å just before the cap structure. The base of this complex, which is open to the cytosol, has an opening of 55Å while the cap, open to the extracellular space, has an opening of 20Å (Fronzes *et al*, 2009). The extracellular T-pilus is composed primarily by VirB2 with VirB7 and VirB5 playing secondary roles (Christie & Cascales, 2005). More specifically, VirB5 may contribute to the protein-protein interactions required for pilus assembly in addition to host cell recognition and possibly even direct cell-to-cell contact (Backert, Fronzes & Waksman, 2008). These pili vary from long and flexible to, as in our example of the *A. tumefaciens* T4SS, short and rigid. All morphological variations aside, it is believed that all conjugative pili are composed of a single pilin subunit that forms a helical filament with a varying internal diameter (Cascales & Christie, 2003).

Mechanism of Transport

The VirB2 and VirB5 subunits which make up the T-pilus are thought to initiate direct cell-to-cell contact before translocation can occur. VirB5 located at the tip of the pilus may even interact with specific receptors displayed on the host cell surface (Backert, Fronzes & Waksman, 2008). Transport via the T4SS is directed by a loosely defined C-terminal signal sequence which can vary between species. For example, *A. tumefaciens*' signal peptide contains arginine clusters while that of *L. pneumophilia* contains several hydrophobic residues (Christie & Cascales, 2005). The coupling proteins (CP) of type IV translocators, VirD4 in the case of *A. tumefaciens*, share sequence similarity with known DNA translocases. This similarity, in conjunction with their DNA, effector protein and ATP-binding ability, suggests a substrate recognition and recruitment role for these proteins (Cascales & Christie, 2003). VirD4 binds and subsequently transfers its substrate to VirB11, another inner membrane ATPase. While both these protein subunits are part of the energy complex, studies suggest that this transfer does not require any expenditure of energy. However the subsequent transfer of the substrate to inner membrane VirB6 and VirB8 subunits, does require coordinate ATP binding and/or hydrolysis by VirD4, VirB4 and VirB11. Also, both VirB6 and VirB8 must be co-synthesized to successfully mediate further translocation of the substrate. Mapping of the translocation pathway places VirB2 and VirB9, which are periplasmic as well as outer membrane associated proteins, at the very distal end of the path. Similar to VirB6 and VirB8, both VirB2 and VirB9 must be co-synthesized for sufficient contact with the substrate. The substrate does not form any detectable interactions with VirB3, VirB5 or VirB10 throughout its translocation journey however deletion studies show that these subunits are essential to substrate translocation (Christie & Cascales, 2005). The proposed role for VirB10 is that of an energy transducer similar to TonB. VirB10 may transduce inner membrane energy to the periplasmic and outer membrane parts of the translocation pathway where there is no direct energy source. Indeed, VirB10 does undergo a conformational change upon ATP association with the inner membrane ATPases. The current model suggests that the major outer wall component of the double-walled transmembrane chamber, as described previously, is VirB10 while the internal wall is composed of VirB7 and VirB9. It has been proposed that the conformational change in VirB10 causes the opening of the constriction site in the cap to allow the transport of substrates (Fronzes *et al*, 2009).

Summary

- The T4SS is a one-step mechanism of DNA and protein translocation in Gram negative and Gram positive bacteria.
- This system is directly involved in virulence and the spread of antibiotic resistance among bacteria.
- Transport relies on three inner membrane ATPases and VirB10 harnessing of the inner membrane energy.
- The signal peptide for this system is located at the C-terminus but is poorly defined and varies between species.
- The exact architecture and mechanism of transport have not been completely elucidated.
- Translocated effector proteins reconstruct their host cell to benefit the invading pathogen (morphological changes, intracellular signaling interference, and nutrient synthesis).
- T4SS-containing bacteria cause a broad spectrum of infection.

Type V Secretion System (T5ss)

Three separate pathways can be categorized as type V secretion systems: the autotransporters (V_a), two-partner secretion system (V_b), and the oligomeric coiled coil adhesin (Oca) family (V_c). These pathways, found only in Gram negative bacteria, facilitate the transport of proteins within the periplasmic space across the outer membrane. As mentioned earlier, these pathways are considered terminal branches of the Sec and Tat pathways which only transport their substrates as far as the periplasmic space. Since the periplasm contains no ATP or proton gradient, outer membrane transport must overcome the lack of a direct energy source. This secretion system is defined by two characteristics; transport of the substrate across the outer membrane via a transmembrane β-barrel pore and a substrate which contains all the information required for transmembrane export (Desvaux, Parham & Henderson, 2004). The type V system contains the simplest outer membrane secretion apparatuses (Figure 4.5) and also represents the largest family of outer membrane protein translocators. Protein precursors of this system can be identified by three domains including the N-terminal signal peptide for inner membrane transport, the extracellular functional or passenger domain (mature protein) and the C-terminal helper domain essential to outer membrane transport. All the mature proteins of the T5SS identified to date, although very diverse in sequence and function, have been linked to bacterial virulence. The roles of type V secreted proteins consist of direct enzymatic activity, cytotoxicity, adhesive ability, bacterial motility, immunomodulation, and the maturation of further virulence factors (Desvaux, Parham & Henderson, 2004).

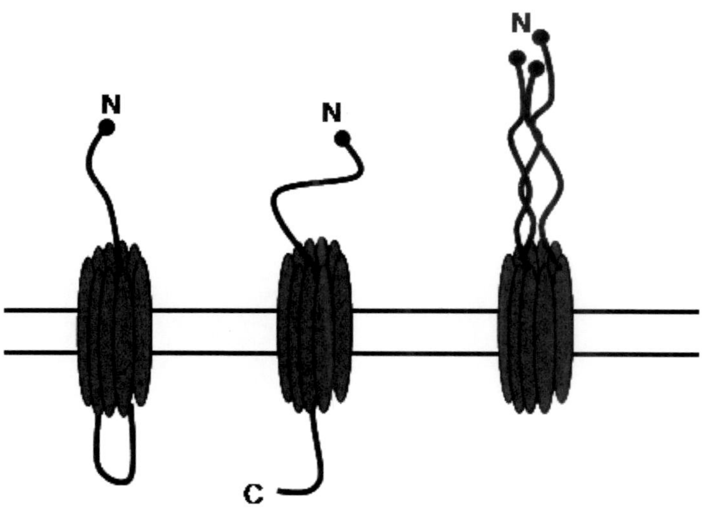

Figure 4.5. Mechanisms of type V outer membrane transport. Autotransporters (Va) insert their C-terminal into the outer membrane to form a β-barrel, the linker region then forms a hairpin and the exoprotein is secreted. Two-partner secretion (Vb) involves two separate polypeptides with one forming the outer membrane pore and the second being the transported exoprotein. The oligomeric coiled-coil adhesins (Vc) have short C-terminals which trimerize to form the outer membrane β-barrel, all three exoproteins are expressed at the surface in a "lollipop" configuration.

Autotransporters (V_a)

The autotransporter (AT) system is the most studied pathway of the T5SS. The protein substrates of this pathway are synthesized as a single polypeptide with three functional domains and contain all the information necessary for secretion. During transport across the inner membrane, most likely by the Sec translocon, the N-terminal signal peptide is cleaved releasing the protein/pre-protein into the periplasmic space. Once in the periplasm, at least one β-strand of the C-terminus inserts into the outer membrane leaving the passenger domain extended in the periplasm (Henderson *et al*, 2004). This C-terminal helper domain is the most conserved feature of autotransporters, usually being between 250-300 residues in length. The predicted folding of these domains is a 12-14 anti-parallel strand β-barrel with alternating hydrophobic side chains that extend into the membrane and hydrophilic side chains projecting into the aqueous interior of the barrel (Thanassi *et al*, 2005). The α-helical linker region (21-39 amino acids) upstream of the β-core is thought to form a hair-pin structure bringing the N-terminal passenger domain up through the newly formed outer membrane pore (Desvaux, Parham & Henderson, 2004). The α-helix would then be responsible for plugging the pore until translocation is complete. This model has been supported by the crystal structure of an *N. meningitidis* type V translocator however there are theories of multimeric pore formation with the protein being translocated *between* the β-barrels (Thanassi et al, 2005). This discrepancy may simply be due to differences among different autotransporters.

Some autotransporters contain an intramolecular chaperone within their passenger domain that ensures the proper folding of the secreted protein. The folding of autotransporter proteins has been suggested to begin in the periplasm prior to or simultaneously with its OM translocation. However, as the outer membrane translocation of these proteins is directed solely by their own β-barrel forming C-terminus, it is unlikely that a large fully folded protein could be handled by this pore. Instead it has been proposed that small periplasmic chaperones maintain the pre-protein in an unfolded and translocation compliant state so that it may pass through the outer membrane and fold on the bacterial surface (Thanassi *et al*, 2005). Again these differences may reflect the folding energies of different substrates, with unfolded polypeptides traveling through a single β-barrel and larger partially folded proteins travelling through an oligomeric structure. In fact, it has been proposed that the energy of an unfolded or partially folded protein is what helps drive the translocation of the protein across the outer membrane (Thanassi et al, 2005).

Once the substrate has transversed the outer membrane it can undergo one of several processing fates. It may be released into the extracellular milieu (via a membrane bound protease or autoproteolysis) or it may remain associated with the membrane, either through a noncovalent interaction with its β-core after cleavage or as an intact transmembrane protein (Desvaux, Parham & Henderson, 2004). The fate of the cleaved β-barrel is not so clear. It seems likely that the β-barrel segment of these proteins is a stable structure as compared with common outer membrane porins of similar structure. Indeed cellular localization studies do show a persistence of these structures within the outer membrane after cleavage of the substrate (Henderson *et al*, 2004). It hardly seems favorable however, for a bacterium to accumulate so many holes within its protective outer layer.

Two-Partner Secretion (V_b)

Two-partner secretion (TPS) shares several similarities with the AT pathway with the main difference being that TPS involves two separate proteins, the secreted exoprotein (TpsA) and the outer membrane transporter (TspB). The genes for these two proteins are most often found within the same operon and are synthesized with N-terminal signal peptides to direct their inner membrane transport via the Sec system (Thanassi et al, 2005). Once the proteins have been exported to into the periplasm, the TspB forms an outer membrane β-barrel much like the AT, followed by the insertion of the passenger domain (TspA) into the outer membrane pore. The β-barrel of TPS is predicted to contain approximately 19-22 antiparallel strands as compared to the 12-14 stranded autotransporters and also possesses a higher level of complexity as it is directly involved in the maturation of the exoprotein (Henderson *et al*, 2004). A conserved N-terminal 110-residue TPS domain found within the exoprotein specifically interacts with the TspB and is responsible for directing the outer membrane transport. Association between the N-terminus of TspA and the TspB pore may result in the opening of the pore which then closes again after the exoprotein has been secreted (Thanassi et al, 2005).

Studies suggest that the TspA passenger domain likely transits the periplasm very rapidly, almost directly bridging inner membrane transport with outer membrane secretion (Thanassi *et al*, 2005). This proposed mechanism seems likely, as many proteins exported by this system are large (>100kDa) and at higher risk of aggregation and degradation within the periplasm. It has also been suggested that the exoprotein folds progressively at the cell surface as it is being translocated, thereby using the energy found within the large unfolded polypeptide to drive secretion (Henderson *et al*, 2004). Thus, considering the requirement for inner membrane transport to be coupled to outer membrane transport in addition to the large substrate size, a periplasmic intermediate is unlikely. Indeed in deletion studies in which the TspB pore protein is removed, the cognate TspA exoprotein becomes trapped within the periplasm and is very sensitive to proteolysis whereas the secreted form exhibits no such sensitivity. Also, if the translation of the TspB transporter is delayed, the exoprotein loses its secretion ability (Jacob-Dubuisson *et al*, 2001). Therefore an uninterrupted pathway from the inner to outer membrane would allow efficient harnessing of the folding energy while at the same time protecting the aggregate-prone polypeptide.

There is also a possibility of C-terminal intramolecular chaperones, similar to that of the autotransporters, which aid in the prevention of premature folding (Jacob-Dubuisson *et al*, 2001). Several of these exoproteins undergo proteolytic processing on the extracellular face of the membrane concomitant with their translocation. At the bacterial surface, the N-terminal TPS domain is responsible for nucleating the folding of the passenger domain Once translocation and folding is complete, the mature exoprotein may remain tethered to the bacterial surface via its C-terminus or may be released through a proteolytic (possibly autoproteolytic) cleavage (Thanassi *et al*, 2005).

Oligomeric Coiled-Coil Adhesins (V_c)

Proteins of this family have only recently been classified as surface attached oligomeric autotransporters. The prototypical YadA of *Yersinia pestis* contains three domains in addition to the N-terminal signal peptide, passenger domain and C-terminal helper region consisting of only four β-strands (Henderson *et al*, 2004). The C-terminal is absolutely essential for outer membrane insertion while a linker region is necessary to maintain the integrity of the entire protein. This system is hypothesized to trimerize within the outer membrane to form a 12-stranded β-barrel similar to that described for the TolC outer membrane factor of type I secretion. The exoproteins of the trimer are displayed at the bacterial surface in what is described as a lollipop-like configuration (Henderson *et al*, 2004).

Summary

- The T5SS is composed of three different pathways: Autotransporters, Two-partner Secretion and Oligomeric Coiled-coil Adhesins.
- This secretion system is found only in Gram negative bacteria as it is a form of outer membrane transport.
- All substrates identified to date are virulence factors.
- Outer membrane transport proposes a unique challenge as there is no ATP available to drive transport.
- The polypeptide substrates of these systems contain all the information required for transport and direct their own translocation with no requirement for other transport proteins.
- Substrates of this system are always translocated via a transmembrane β-barrel formed by the C-terminal helper/translocation domain of the substrate.
- Several substrates possess intramolecular chaperones in addition to exhibiting autoproteolysis ability.

Type VI Secretion System (T6ss)

The type VI secretion system is a newly characterized protein transport system of Gram negative bacteria. This system is widespread, found in both pathogenic and non-pathogenic species, however in some of those pathogenic species the T6SS has been found to be a key virulence factor. The type VI system of *Vibrio cholera* for example, is known to translocate potential effectors into eukaryotic host cells. In fact, the T6SS was first defined in *V. cholera* with the identification of a hemolysin co-regulated protein Hcp and three related VgrG proteins that were secreted without the cleavage of a signal sequence, unlike other secreted proteins of *V. cholera* (Bingle, Bailey & Pallen, 2008). In addition, it has also been noted that the secretion of Hcp and VgrG proteins are co-dependent, neither protein is successfully secreted in the absence of the other (Pukatzi, McAuley & Miyata, 2009). Noting that all the identifiable exported proteins lacked signal peptides and that *V. cholera* did not contain type

III or IV secretion systems, it was concluded in 2006 that there must be another secretory system. This new system was named VAS (virulence associated secretion) and was subsequently identified in *Pseudomonas aeruginosa*, *Escherichia coli* and *Aeromonas hydrophila* among others (Bingle, Bailey & Pallen, 2008). This secretion system is regulated at both the transcriptional and post-transcriptional levels. The T6SS gene clusters identified thus far encode 12 to 25 proteins. Among these proteins are ATPases, putative outer membrane lipoproteins and chaperones (Filloux, Hachani & Bleves, 2008).

Architecture and Mechanism of Transport

The Hcp protein mentioned earlier forms hexameric rings of 40Å in diameter and 100nm in length and is secreted by all T6SS-containing bacteria. As such, it is suggested that Hcp is a secreted structural component acting as the conduit for export across the cell envelope (Pukatzi, McAuley & Miyata, 2009). In addition, the VgrG proteins share significant structural similarity with the membrane puncturing needle complex of T4 bacteriophages and, like that of the T4's, form multimeric complexes. It is unclear whether this structure is used to break through the bacterial membrane, the host membrane, or both. The VgrG proteins also contain catalytic domains which suggest that they play another role in addition to the formation of the type VI secretion apparatus (Filloux, Hachani & Bleves, 2008).

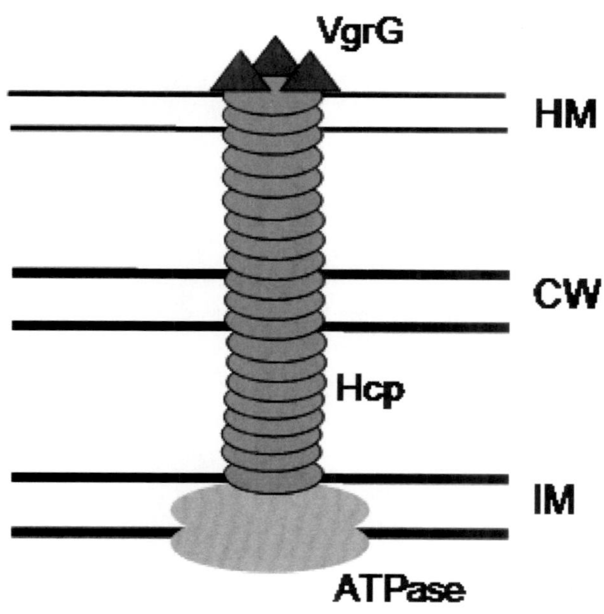

Figure 4.6. Simplified schematic of the T6SS. ATPases located within the inner membrane (IM) provide the energy necessary for secretion. The hexameric Hcp rings form a channel structure which is topped by a trimeric VgrG complex much like the cell-puncturing device of a T4 bacteriophage. This VgrG complex may be responsible for puncturing the cell wall (CW) and/or the host cell membrane (HM).

One proposed mechanism involves the localization of the T6SS subunits within the inner membrane, followed by the export and assembly of Hcp and VgrG subunits in the periplasm. At this point the VgrG proteins may form a trimeric complex resembling the cell-puncturing complex of T4s, while the hexameric Hcp rings dock beneath (Figure 4.6). As rings are continuously added, the tube grows and the tip may either exit through a pore in the cell wall or simply force its way out. If a host cell is within close proximity, the tip of the T6SS apparatus may puncture the cell's membrane and the C-terminal effector domain of the VgrG complex unfolds and interacts with targets within the host cytosol. Alternatively, the VgrG complex may completely detach within the host cell, leaving an open passageway between host and bacterium, or the C-terminus may be cleaved to interact with targets distant from the puncture site (Pukatzi, McAuley & Miyata, 2009).

Summary

- Type VI exported proteins contain no signal sequence.
- Transport is likely driven by ATP hydrolysis.
- The T6SS is only found in Gram negative bacteria, both pathogenic and non-pathogenic.
- The type VI apparatus contains a trimeric VgrG complex which bears resemblance to the cell-puncturing tool of a T4 bacteriophage.
- Transport is likely directly across the bacterial envelope and the host cell membrane.
- Regulation occurs transcriptionally as well as post-transcriptionally.

Type VII Secretion System (T7ss)

The recently identified T7SS is also known as the ESX-1 system as it is responsible for the secretion of prototypic ESX proteins. The ESX-1 system of *Mycobacterium tuberculosis* is involved in the secretion of the 6kDa early secreted antigenic target (ESAT-6) and the 10kDa culture filtrate protein (CFP-10) which together form a 1:1 heterodimer important in host interactions. Interestingly, this system is absent in the attenuated strains *Mycobacterium bovis* BCG and *Mycobacterium microti*, signifying the involvement of T7SS in pathogenesis. However, while the ESTAT-6 and CFP-10 proteins are important to host-pathogen interactions these proteins have been found in supernatants of *in vitro* grown *M. tuberculosis*, implying that the secretion of these products is not dependent on direct interaction with a host. This observation differs from the other secretion systems of pathogenic bacteria responsible for releasing virulence factors and suggests that while these proteins are important to pathogenesis, they may also serve a more general purpose for the bacterium (Simeone *et al*, 2009). Genomic analysis has also predicted the presence of type VII secreted proteins in a range of actinobacteria and Gram-positive bacteria, identifiable by their approximate size of 100 amino acids and a C-terminal WXG motif. The mechanism of secretion is proposed to resemble that of the T4SS discussed earlier. It is suggested that two ATPases, one membrane bound and one cytosolic, direct the substrate across the cytoplasmic membrane and fuel the process by ATP hydrolysis (DiGiuseppe Champion *et al*, 2006). Effectors translocated by the T7SS are reported to induce a strong immune response mediated by T cells, cause necrosis, apoptosis, membrane lysis and cytolysis as well as suppress proinflammatory responses. The type VII system has also been implicated in the recently described translocation of *M. tuberculosis* from the phagosome into the host cell cytoplasm (Simeone et al, 2009).

Architecture and Mechanism of Transport

As a recently characterized transport system, the exact mechanism and architecture of the secretion apparatus remains to be elucidated. Several genes have been found to be required for transport through mutation and knockout studies, however for the most part their functions remain unknown (DiGiuseppe Champion & Cox, 2007). What is clear, is that this system is very complex and relies on multiple protein complexes for successful translocation of substrates to the external milieu. DiGiueseppe Champion and others (2006) do propose a model for secretion that is similar to that of type IV secretion. It is suggested that ESAT-6 and CFP-10 fold within the cytoplasm to form a stable dimer and are then recognized by the cytoplasmic Rv3871 via the C-terminal signal of the CFP-10 subunit. The complex is then transported to membrane bound Rv3870 which works, along with Rv3871, to transport the substrate across the cellular envelope possibly through Rv3877 (Figure 4.7). It remains unclear however, how the secreted protein negotiates the cell wall and lipid layers once across the cytoplasmic membrane. Currently there is no known mycobacterial secretion machinery found outside the cytosol (DiGiuseppe Champion & Cox, 2007).

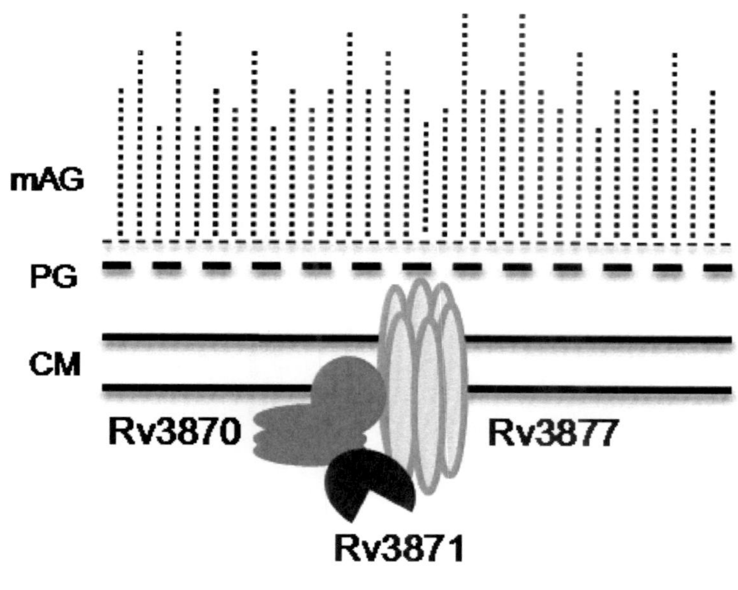

Figure 4.7. The proposed type VII secretion system of Mycobacterium tuberculosis. Cytoplasmic Rv3871 identifies target substrates via a C-terminal signal sequence and directs substrates to membrane bound Rv3870 and together these two ATPase transporter subunits direct translocation across the cytoplasmic membrane. The mechanism of translocation across the peptidoglycan (PG) cell wall and mycolyl-arabinogalactan (mAG) lipids layers remains unknown.

Summary

- The T7SS is a complex system involving several components, all of which have to yet to be defined.
- Proteins are likely transported in a folded state and transport is driven by ATP hydrolysis.
- Substrates are often ~100 aa immunogenic proteins and translocation is dependent on a disordered C-terminal sequence containing the WXG motif.
- This type VII system is likely associated with the pathogenesis of *M. tuberculosis*, though may also serve a more general purpose.
- This system may aid in *M. tuberculosis'* escape from the phagosome into host cell cytoplasm.

Conclusion

Bacteria have evolved several ways to export proteins across their protective barriers and many of these systems are directly linked with virulence. It is important to note that individual bacterial species often express more than one secretion system to carry out their secretory needs. Secretion systems I though VII (summarized in Figure 4.8) demonstrate the mechanistic variety among protein export machines. Although mechanistically these systems can differ greatly, they seem to share some significant structural and architechetural similarities. Interestingly, however, research strongly suggests that these transporters all evolved independent of one another (Saier 2006).

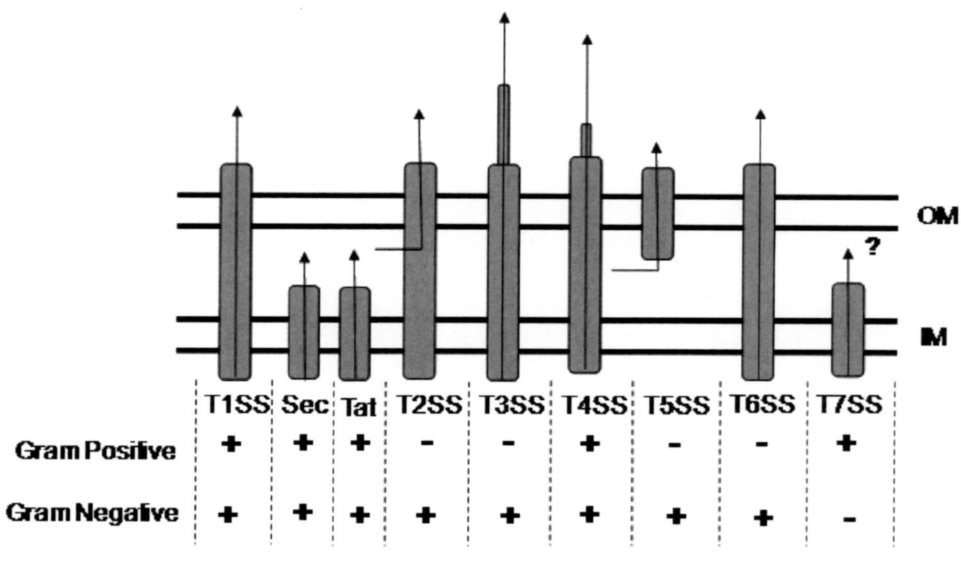

Figure 4.8. A summary of secretion systems type I-VII.

References

Alami, M, S Deitermann, G Eisner, H G Koch, J Brunner, and M Muller. "Differential interactions between a twin-arginine signal peptide and its translocase in Escherichia coli." *Molecular Cell*, 2003: 937-46.

Backert, S, R Fronzes, and G Waksman. "VirB2 and VirB5 proteins: specialized adhesins in bacterial type-IV secretion systems?" *Trends in Microbiology*, 2008: 409-13.

Berks, B C, T Palmer, and F Sargent. "Protein targeting by the bacterial twin-arginine translocation (Tat) pathway." *Current Opinion in Microbiology*, 2005: 174-81.

Bingle, L E, C M Bailey, and M J Pallen. "Type VI secretion: a beginner's guide." *Current Opinion in Microbiology*, 2008: 3-8.

Cascales, E, and P J Christie. "Definition of a Bacterial Type IV Secretion Pathway for a DNA Substrate." *Science*, 2004: 1170-73.

Cascales, E, and P J Christie. "The versatile bacterial type IV secretion systems." *Nature Reviews Microbiology*, 2003: 137-49.

Christie, P J, and E Cascales. "Structural and dynamic properties of bacterial Type IV secretion systems." *Molecular Membrane Biology*, 2005: 51-61.

Cianciotto, N P. "Many substrates and functions of type II secretion: lessons learned from Legionella pneumophila." *Future Microbiology*, 2004: 797-805.

Coburn, B, I Sekirov, and B B Finlay. "Type III secretion systems and disease." *Clinical Microbiology Review*, 2007: 535-49.

Cornelis, G R. "The type III secretion injectisome." *Nature Review Microbiology*, 2006: 811-25.

De Buck, Emmy, Elke Lammertyn, and Jozef Anne. "The importance of the twin-arginine translocation pathway for bacterial virulence." *Trends in Microbiology*, 2008: 442-53.

Dekker, C, B DeKruijff, and P Gros. "Crystal structure of SecB from Escherichia coli." *Journal of Structural Biology*, 2003: 313-19.

Delepelaire, P. "Type I Secretion in Gram Negative Bacteria." *Biochimica at Biophysica Acta*, 2004: 149-161.

Desvaux, M, N J Parham, and I R Henderson. "Type V Protein Secretion: Simplicity Gone Awry?" *Current Issues in Molecular Biology*, 2004: 111-24.

DiGiuseppe Champion, P A, and J S Cox. "Protein systems in Mycobacteria." *Cellular Microbiology*, 2007: 1376-84.

DiGiuseppe Champion, Patricia A, Sarah A Stanley, Matthew M Champion, Eric J Brown, and Jeffery S Cox. "C-Terminal Signal Sequence Promotes Virulence Factor Secretion in Mycobacterium tuberculosis." *Science*, 2006: 1632-36.

Filloux, A. "The underlying mechanism of type II protein secretion." *Biochimica et Biophysica Acta*, 2004: 163-79.

Filloux, A, A Hachani, and S Bleves. "The bacterial type VI secretion machine: yet another player for protein transport across membranes." *Microbiology*, 2008: 1570-83.

Flieger, O, A Engling, R Bucala, H Lue, W Nickel, and J Berhagen. "Regulated secretion of macrophage migration inhibitory factor is mediated by a non-classical pathway involving an ABC transporter." *FEBS Letters*, 2003: 78-86.

Fronzes, R, E Schafer, L Wang, H R Saibil, E V Orlova, and G Waksman. "Structure of a Type IV Secretion System Core Complex." *Science*, 2009: 266-68.

Galan, J E. "Energizing type III secretion machines: what is the fuel?" *Nature Structural and Molecular Biology*, 2008: 127-8.

Galan, J E, and A Collmer. "Type III secretion machines: bacterial devices for protein delivery into host cells." *Science*, 1999: 1322-8.

Henderson, I R, F Navarro-Garcia, M Desvaux, R C Fernandez, and D Ala'Aldeen. "Type V Protein Secretion Pathway: the Autotransporter Story." *Microbiology and Molecular Biology Reviews*, 2004: 692-744.

Holland, I B, L Schmitt, and J Young. "Type 1 protein secretion in bacteria, the ABC-transporter dependent pathway." *Molecular Membrane Biology*, 2005: 29-39.

Jacob-Dubuisson, F, C Locht, and R Antoine. "Two-partner secretion in Gram-negative bacteria: a thrifty, specific pathway for large virulence proteins." *Molecular Microbiology*, 2001: 306-313.

Journet, L, K T Hughes, and G R Cornelis. "Type III secretion: a secrtory pathway serving both motility and virulence." *Molecular Membrane Biology*, 2005: 41-50.

Juhas, M, et al. "Novel Type IV Secretion System Involved in Propagation of Genomic Islands." *Journal of Bacteriology*, 2007: 761-71.

Kim, J, and D A Kendall. "Sec-Dependent Protein Export and the Involvment of the Molecular Chaperone SecB." *Cell Stress Chaperones*, 2000: 267-75.

Koronakis, V, J Eswaran, and C Hughes. "Structure and Function of TolC: The Bacterial Exit Duct for Proteins and Drugs." *Annual Review of Biochemistry*, 2004: 467-89.

Missiakas, D, and S Raina. "Protein Folding in the Bacerial Periplasm." *Journal of Bacteriology*, 1997: 2465-71.

Mori, H, and K Ito. "The Sec protein translocation pathway." *Trends in Microbiology*, 2001: 494-500.

Muller, M. "Twin-arginine-specific protein export in Escherichia coli." *Research in Microbiology*, 2005: 131-36.

Pukatzi, S, S B McAuley, and S T Miyata. "The type VI secretion system: translocation of effectors and effector-domains." *Current Opinion in Microbiology*, 2009: 11-17.

Robinson, C, and A Bolhuis. "Protein targeting by the twin-arginine translocation pathway." *Nature Review Molecular Cell Biology*, 2001: 350-56.

Saier, M H. "Protein secretion and membrane insertion systems in Gram-negative bacteria." *Journal of Membrane Biology*, 2006: 75-90.

Schroder, G, and C Dehio. "Virulence-associated type IV secretion sytems of Bartonella." *Trends in Microbiology*, 2005: 336-42.

Simeone, R, D Bottai, and R Brosch. "ESX/type VII secretion systems and their role in host-pathogen interaction." *Current Opinion in Microbiology*, 2009: 4-10.

Sklar, J G, T Wu, D Kahne, and T J Silhavy. "Defining the roles of the periplasmic chaperones SurA, Skp, and DegP in Escherichia coli." *Genes & Development*, 2007: 2473-84.

Stathopoulos, C, D R Hendrixson, D G Thanassi, S J Hultgren, J W St. Geme, and R Curtiss. "Secretion of virulence determinants by the general secretory pathway in Gram-negative pathogens: An evolving story." *Microbes and Infection*, 2000 : 1061-72.

Stephenson, K. "Sec-dependent protein translocation across biological membranes: evolutionary conservation of an essentail protein transport pathway." *Molecular Membrane Biology*, 2005: 17-28.

Thanassi, D G, C Stathopoulos, A Karkal, and H Li. "Protein secretion in the absence of ATP: the autotransporter, two-partner secretion and chaperone/usher pathways of Gram-negative bacteria." *Molecular Membrane Biology*, 2005: 63-72.

Tjalsma, H, et al. "The Role of Lipoprotein Processing by Signal Peptidase II in the Gram-positive Eubacterium Bacillus subtilis." *The Journal of Biological Chemistry*, 1999: 1698-1707.

Tuteja, R. "Type I signal peptidase: An overview." *Archives of Biochemistry and Biophysics*, 2005: 107-11.

In: Molecular Aspects of Infectious Diseases
Editor: Mazen T. Saleh

ISBN: 978-1-61728-690-2

Chapter V

Viruses

Mazen Saleh
Department of Biology and Northen Ontario School of Medicine, Laurentian University, Sudbury, Ontario, Canada, P3E 2C6

Viruses may be the most abundant microbes on our planet. Fortunately for us the vast majority of them are present in the oceans and they infect other microbes. Some however can be found in fresh water systems and find their way into our potable waters. The most common are the rotaviruses that cause diarrhea and hepatitis viruses that cause liver cirrhosis and liver cancer. The most common viral infections are those of the respiratory system that cause respiratory syndromes, including the common flu, and more involved viral pneumonias. We all recall the Spanish flu of 1912 that claimed the lives of over 20 million people.

Viruses are included in the term "microbes" simply because of their small size. Although, the largest of the viruses are comparable in size to the smallest of the Bacteria and the Archae. An important difference between bacteria and viruses, one which complicates the treatment of viral infections, is that viruses are non-living entities and can not multiply without a host. Another factor that contributes to the difficulty in the treatment of viral infections is their simplicity in structure. This provides but a very limited selection of targets for antivirals.

Basic Structure

In their simplest form, viruses are made up of proteinaceous enclosed vessel carrying within the genetic material or genome of the virus. This vessel is known as a capsid and can take on a variety of shapes. The common shape is an icosohedral form produced by the self assembly of a single protein molecule or homologous subunits (Figure 5.1). Viruses exist where this simple form becomes more complex by having enzymes and other molecules within the capsid and by having an envelope around the capsid (Figure 5.1). This creates an additional compartment between the surface of the capsid and the envelop. The envelop is a typical lipid bilayer containing membrane proteins. The enzymes carried within the capsid serve crucial functions for the virus during host infection and are required for its competence and survival. The genome of the virus may also be one of several forms; it can be DNA or RNA in structure, and it may be single or double stranded in form. Single stranded RNA genomes may further vary in either being sense single stranded RNA genomes (+ssRNA) or antisense single stranded RNA genomes (–ssRNA). As a result, viral replication cycles within host cells are dictated by the structure of their genomes. The basic requirement for successful replication within the host is the production of the two basic components of the viral particle, the genome and the proteins that make up the capsid and other structures or enzymes. Because of this, we find that certain viruses may infect host cells and replicate without invading the nucleus of the cell while others must penetrate the nucleus to replicate successfully (Figure 5.2).

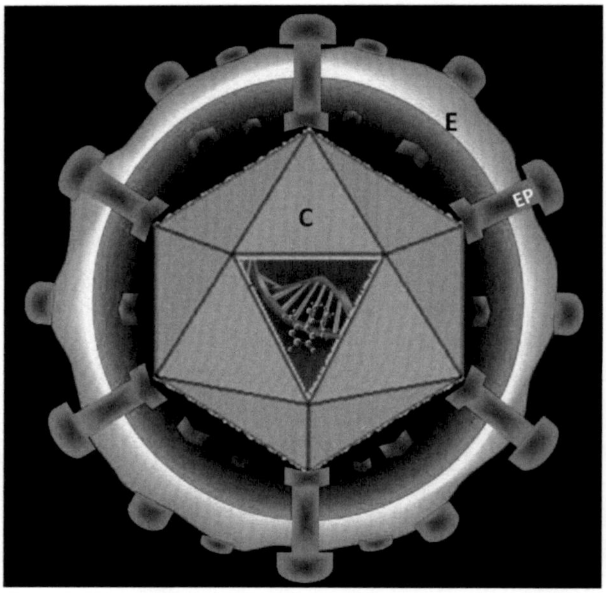

Figure 5.1. Schematic of a typical virus structure. In this case, the capsid (C) is shown in icosahedral geometry and enclosed within an envelop (E). Spikes of surface adhesins, enzyme, and other envelop proteins (EP) are shown within the membrane. Contained within the capsid is the viral genome with associated proteins in certain cases.

The Viral Lifecycle within Host Cells

As an example to highlight the key events and molecular interactions during host cell infection the case of the human immunodeficiency virus (HIV) will be presented. The HIV is known as a retrovirus. Retroviruses are named as such because during infection and multiplication within host cells genetic information flows in the reverse direction as compared to the normal direction of flow. Genetic information normally flows as follows: DNA → RNA (gene transcription) → protein (gene translation). During retroviral infections genetic information flows as follows: RNA (viral genome) → DNA → RNA (gene transcription) → protein (gene translation). The key feature of these events is the transformation of the RNA genome of the virus to a DNA genome. Once in the form of DNA, the genome may integrate into the genome of the infected cell and remain in that form for extended periods of time, often years. This produces a latent infection. Under suitable conditions, the "provirus" is activated and removes itself from the genome of the host to be transformed into the original RNA form (Figure 5.3).

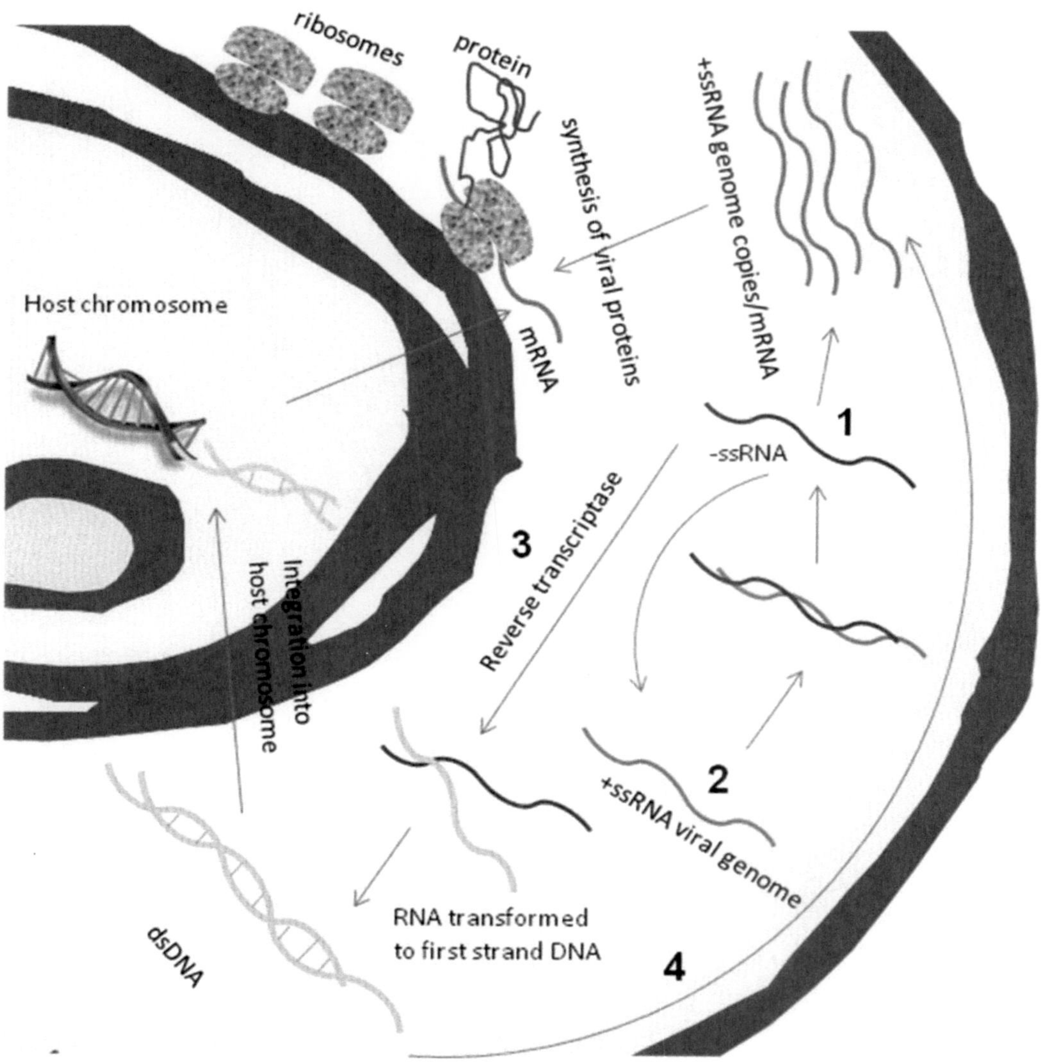

Figure 5.2. Multiplication of viral genomes in infected cells. Depending on the nature of viral genome, replication may take place in the cytoplasm or the nucleus. The various transformations of viral genomes shown represent the required transformations for successful replication. For example, a virus with a –ssRNA (1) must produce a complementary ssRNA to be used as a template for further production of –ssRNA (actual replication of genome). The reverse applies to viruses with +ssRNA genomes (2). Retroviruses achieve replication and shromosome insertion by transformin their genomes from RNA to DNA via the activity of reverse transcriptase (3) and produce viral proteins by transcription or +ssRNA production (4).

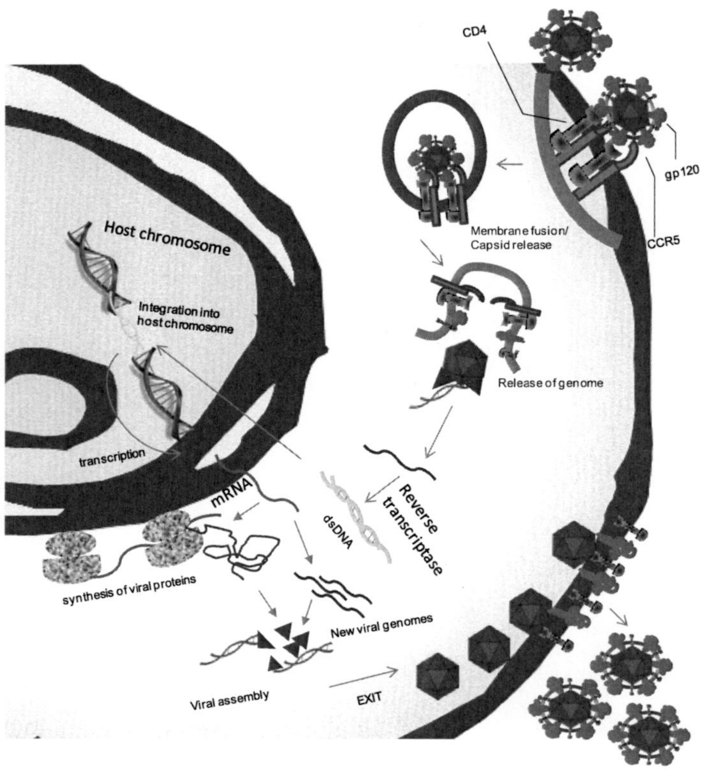

Figure 5.3. Schematic of HIV attachment to host cell, cell penetration, replication, and release.

Survey of Medically Important Viruses

As viruses are non-living entities they are not capable of secreting toxins as with bacteria and other microbes. Nor can they form such dormant structures as bacterial spores, which can survive outside the host under fluctuating environmental conditions for extended periods of time. Formation of spores is not only an effective survival strategy for bacteria and other pathogens but is an efficient method of transmission, as for example the case with inhalation anthrax. As a result, pathologies associated with viral infections are the result of direct action of viruses on host tissues. This includes direct effects on the target tissue infected by the virus as well as indirect effects on tissue due to the response of the host immune system. The tissue that the virus infects is determined mostly by the viral adhesins and their cognate receptors on specific host cells. This is important to remember because infection of different tissues by the same virus will produce different pathologies. It is also important to remember that pathogen-host interaction is specific in nature. In other words, the interactions between the surface of a pathogen and the surface of host cells are mediated by specific molecular interactions. This fact is exemplified by the observation that the Northern European population is more resistant to HIV infections. This resistance correlates with deletion variants of the co-receptors for the virus called CCR5 and CXCR4 (1-5).

Table 5.1. Summary of medically important viral infections.

Virus	Genome	Envelop	Target tissue	Disease	Key Features
Influenza	–ssRNA	yes	Respiratory epithelium	Influenza/flu	Hypervariable surface antigens. Immunization difficult.
Coronavirus	+ssRNA	yes	Respiratory epithelium	SARS	Involvment of the lower respiratory system and pneumonia. Can be fatal in healthy individuals.
HIV	+ssRNA	yes	Immune tissue, T-cells, macro-phages, dendritic cells and monocytes	AIDS	HIV is a retrovirus and produces a latent infection. Difficult to target as a provirus. The reverse transcriptase enzyme it carries within the capsid can be targeted with antivirals. Attack of immune tissue weakens the immune system and increases the incidence of secondary bacterial and fungal infections in the respiratory system.
West Nile	+ssRNA	yes	Various epithelia, systemic infection, including the meninges	West Nile fever	Mild flu-like systems. In about 1% of those infected meningitis and encephalitis may develop.
Varicella-Zoster	dsDNA	yes	Respiratory epithelium and neurons	Chickenpox and Shingles	Following the appearance on various parts of the skin, the virus may exist in a dormant form in sensory neurons. At a later stage, typically years or decades, the virus moves from the sensory neurons to the skin to cause Shingles.
Variola	dsDNA	yes	Respiratory epithelium	Smallpox	Sudden fever and formation of skin rashes that develop into pustules. Can be fatal in over 20% of cases
Rubeola/ Measles	–ssRNA	yes	Respiratory epithelium and immune tissue	Measles	Cough and fever present along with skin eruptions. Koplik's spots in mouth
Polio	+ssRNA	No	Mucosa of the oral cavity and the GI tract	Poliomyelitis	Invasion of mucosa and development of viremia. Full recovery in most cases. In tohers, the virus enters the central nervous system and causes paralytic polio
Ebola and Marburg	–ssRNA	yes	Endotheliu m tissue	Viral hemorrhagic fever	Rapid onset of fever and whole body aches. Attack and destruction of the vascular system leading to whole system hemorrhage and death in over 70% of cases
Hantavirus	–ssRNA	yes	Respiratory epithelium	Hantavirus pulmonary syndrome (HPS)	Rapid onset of symptoms, fever and respiratory complications. Can be fatal.
Rabies/ Lyssavirus	–ssRNA	yes	Salivary glands, muscle tissue, and neurons	Rabies	Zoonotic disease. Unique combinations of symptoms include irritability, anxiety, loss of apatite and fear of water (hydrophobia). Also sensitivity to light and sounds. Those who succumb to the disease show distinct Negri bodies in brain tissue.

Table 5.1 (Continued)					
Rubella	+ssRNA	yes	Respiratory epithelium	German Measles	Skin rash, small red spots and mild fever. Resolves within a few days. Pregnant women in their first trimester are particularly susceptible and exposure to the virus can lead to premature delivery or even death of the fetus. Successful pregnacies often deliver a baby with various syndromes affecting the eyes, the heart, or the ears.
Hepatitis A	+ssRNA	No	Epithelium of the GI tract and liver parenchymal cells	Hepatitis	Also known as infectious hepatitis with acute infections. Spread by fecal contamination of objects, food and water.
Hepatitis B (HBV)	dsDNA	yes	Epithelia, liver parenchymal cells and others	Hepatitis	Spreads sexually, through blood, needles and body secretions. Subclinical acute and chronic infections. May lead to liver cirrhosis and hepatocarcinoma.
Hepatitis C	+ssRNA	yes	Epithelia, liver parenchymal cells and others	Hepatitis	Spreads sexually or through blood. Subclinical acute and chronic infections that may lead to hepatocarcinoma.
Hepatitis D	−ssRNA	Virusoid, RNA only	Epithelia, liver parenchymal cells and others	Hepatitis	Also known as hepatitis delta and is a virusoid. It only co-infects with HBV.
Hepatitis E	+ssRNA	No	Epithelium of the GI tract and liver parenchymal cells	Hepatitis	Produces acute infections. Spread by fecal contamination of objects, food and water.
Hepatitis G	+ssRNA	yes	Epithelia, liver parenchymal cells and others	Hepatitis	Spreads sexually or parenterally. Causes chronic liver inflammation.

A list of the more common viruses that cause human disease is shown in Table 5.1. Some key features of these viruses and the diseases they cause are included to allow for correlation between the structure of the virus, its target host tissue, and the clinical symptoms observed in patients with those infections.

Therapeutic Strategies

As mentioned above, viral infections are more difficult to treat with antibiotics than other microbial infections. This is a direct result of their simplicity in structure and the fact that they multiply within host cells. This obligate intracellular life cycle then shields the virus from both the immune system and from antiviral drugs. From Figure 5.2, it can be seen that the possible processes that can be effectively targeted for the development of antivirals are: (1) the initial adsorption and penetration step; (2) viral genome replication and; (3) viral assembly and release. The most successful antivirals available today are those that target the viral genome replication step. This has been shown to be the case in HIV infections. The strategy followed in this effort is to produce nucleotide analogs which would be incorporated into the newly synthesized viral genome. If this step is accomplished, it will produce deleterious effects on the successful multiplication of the virus as the incorporated nucleotide analogs produce deletions and prevent the synthesis of new viral genomes (Figure 5.4). Other antivirals have been developed that target an enzyme of the flu virus called neuraminidase.

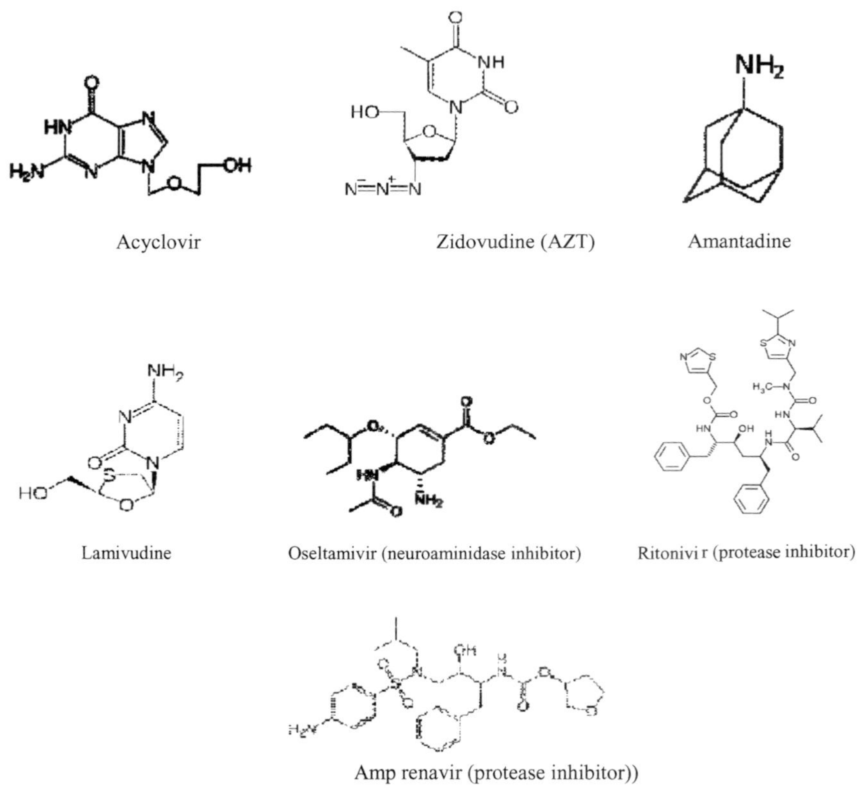

Figure 5.4. Structures of representative antivirals.

This enzyme is a surface molecule of the virus and is required for successful adsorption and penetration of host cells. Another class of antivirals targets viral proteases which are required for processing of capsid proteins and efficient viral assembly before release. These two enzyme classes however are not present in all human viruses.

References

[1] Samson M, Labbe O, Mollereau C, Vassart G, Parmentier M (1996). Molecular cloning and functional expression of a new human CC-chemokine receptor gene. *Biochemistry* 35: 3362–7.

[2] Samson M, Libert F, Doranz BJ, Rucker J, Liesnard C, et al. (1996) Resistance to HIV-1 infection in Caucasian individuals bearing mutant alleles of the CCR-5 chemokine receptor gene. *Nature* 382: 722–725

[3] Anderson J, Akkina R (2007). Complete Knockdown of CCR5 by lentiviral vector-expressed siRNAs and protection of transgenic macrophages against HIV-1 infection. *Gene Therapy* 14: 1287–1297.

[4] a b Galvani AP, Slatkin M (2003). Evaluating plague and smallpox as historical selective pressures for the CCR5-Delta 32 HIV-resistance allele. *Proc. Natl. Acad. Sci. U.S.A.* 100: 15276–9.

[5] Stephens J et al. (1998). Dating the origin of the CCR5-Delta32 AIDS-resistance allele by the coalescence of haplotypes (PDF*). Am J Hum Genet* 62: 1507–15.

[6] Mercer J, Schelhaas M, Helenius A (2010). Virus entry by endocytosis. *Annu Rev Biochem* 79:6.1-6.31

In: Molecular Aspects of Infectious Diseases
Editor: Mazen T. Saleh
ISBN: 978-1-61728-690-2

Chapter VI

The Innate Immune System

Robert M. Lafrenie
Regional Cancer Program, Sudbury Regional Hospital, Sudbury Ontario; Medical Sciences Division, Northern Ontario School of Medicine, Sudbury, Ontario

Abstract

The innate immune system provides the first line of defense against infection by bacteria, viruses, and other pathogens. The cell layers that make up the skin epithelium and mucosa of the respiratory and gastrointestinal tracts provide a barrier to the external environment and form physical, chemical, and biological defenses to counter infections. When these tissues are compromised, anti-pathogenic chemicals and peptides are released and the leukocytes that reside in these tissues recognize and are activated by the infective agents. These surveilling leukocytes which include, granulocytes, macrophages, and dendritic cells have specific cell surface receptors that bind molecules unique to the pathogen. Activation of these cells results in phagocytosis of the pathogens and the release of a variety of bioactive compounds including proteases and reactive oxygen species which can degrade the pathogen or compounds such as cytokines, chemokines, and coagulation proteins that can promote accumulation of additional immune cells. The resulting amplification of the innate immune response generates an inflammatory response where multiple cell types and effectors collaborate to eliminate the infecting pathogen.

1. General Description

The innate immune response is the initial reaction to pathogens or microbes which prevents infection and can eliminate infecting organisms. It is an ancient system found in all classes of plant and animal life. During a normal innate immune response, shared molecular structures expressed on invading pathogens are recognized by host immune cells via a group of pattern recognition receptors that trigger multiple effector mechanisms to eliminate the pathogen (Nurnberger et al., 2004; Akira et al., 2006). In vertebrates, the innate immune response is activated within the first four hours following infection, followed by an induced inflammatory response which can last a few days (Figure 6.1). A subsequent adaptive immune response can be induced over several additional days. The shared molecular structures present on pathogens that initiate an innate immune response are called "pathogen-associated molecular patterns" (PAMPs) (Aderem and Ulevitch, 2000). These molecules are unique to the pathogens and are not expressed by the host organism providing specificity for immune recognition. The PAMPs bind to a small set of genomically-encoded signaling receptors on the host cell, called pattern recognition receptors (PRR) (Medzhitov and Janeway 2002; Kimbrell and Beutler, 2001). This interaction can promote cellular activation or enhance pathogen recognition, resulting in anti-pathogen activity. Thus, the innate immune system responds to infecting agents in a generic or non-specific manner but unlike the adaptive immune system does not confer long term, protective immunity against re-infection with the same pathogens.

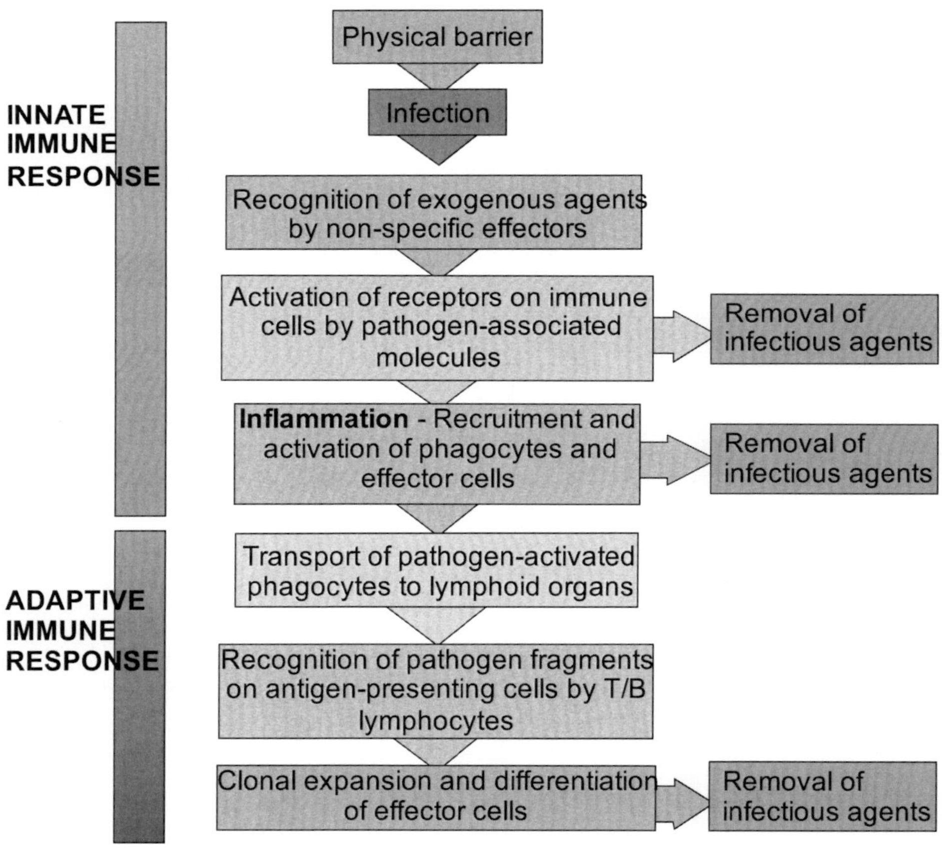

Figure 6.1. The steps in an immune response that show the relationship between innate and adaptive immunity. Important events in an escalating immune response are shown, starting with the formation of an infection following a breach in the physical barrier formed by an epithelium. The immune escalation can be stopped at various points in the procedure by Removal of the infectious agent.

The major functions of the vertebrate innate immune system include the recruitment of immune cells to the site of infection, activation of complement, the clearance of foreign substances, and the activation an adaptive immune response (Akira et al., 2006). Cells of the innate response, such as neutrophils and monocytes, circulate in all tissues (Medzhitov and Biron, 2003). Upon encountering pathogen-specific molecules, these cells are activated to produce and secrete a variety of protein factors including cytokines and chemokines which recruit and activate other immune-related cells (Lin and Karin, 2007). Since neutrophils and monocytes are also capable of phagocytosis and degradation of pathogens, their activation promotes the clearage of the infecting organisms. The presence of pathogens can also activate complement pathways (Gasque, 2004). The complement components are part of a protein cascade which undergoes proteolytic cleavage and activation in response to binding to specific pathogen components. Activation of the complement cascade results in the production of chemokines, factors which promote cell migration to the site of infection, opsonins, factors that promote phagocytosis, and the creation of the complement attack complex which lyses target cells. Monocyte/macrophages that have phagocytosed the pathogens are capable of re-presenting components of the degraded pathogen to T cells resulting in activation of the adaptive immune response (Harding and Song, 1994). Thus, macrophages provide the link between recognition by the innate immune system and activation of the adaptive immune system in response to pathogen infections.

2. Nature of the Innate Immune System Repertoire

The innate immune system can be activated by exposure to a broad range of pathogens even though a relatively small number of specific immune receptors are involved (Ozinsky et al., 2000). Pathogens, especially bacteria, have molecular structures that are not shared by mammalian cells. In general, these structures comprise a relatively small group of invariant, related shapes which are expressed by a wide variety of pathogens and are referred to as "Pathogen-associated molecular patterns" (PAMPs) (Aderem and Ulevitch, 2000). These microbial shapes are recognized during the innate response (Figure 6.2). PAMPs are usually essential actors in microbial survival which means they are not very susceptible to change making them stable targets for identifying pathogens (Akira et al., 2006). For example, bacterial lipopolysaccharide (LPS), also called endotoxin, is a major component of the outer membrane of Gram-negative bacterial cell walls and is a potent activator of monocytes and macrophages (Triantafilou and Triantafilou, 2002). LPS contributes to the structural integrity of bacteria protecting the membrane from some chemical attacks and bacterial death results from mutations or deletions of LPS. Similarly, the peptidoglycan structures that are required to maintain the cell wall integrity of Gram-positive bacteria, such as lipoteichoic acid, are also potent activators of the innate immune system (Yang et al., 2001). Flagellin, the protein subunit of bacterial flagella, is present in high levels in motile bacteria and can promote a strong innate immune response (Hayashi et al., 2001). Mannose-rich glycans derived from microbial glycoproteins, interact with macrophage mannose receptors which promote phagocytosis by macrophages (Sahly et al., 2008). Mannose-rich glycans also interact with mannose-binding lectin in the blood plasma which causes opsonization of bacterial phagocytosis and the activation of the lectin pathway of complement (Turner, 1996). Phosphatidylcholine and related molecules, derived from microbial membranes, can interact with and activate plasma C-reactive protein resulting in opsonization of phagocytosis and complement activation (Fujita et al., 2004).

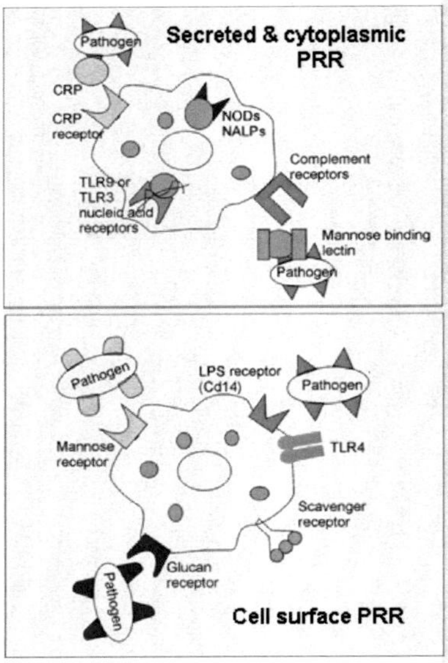

	Receptor	Ligand	Response to activation
Secreted	pentraxin family - C-reactive protein	phosphatidyl serine	promote endocytosis and activate complement
	mannose binding lectin	mannose containing glycoproteins/glycolipids	activate complement
	complement receptors	complement components	promote endocytosis and chemotaxis
Cytoplasmic	NODs	Muramyl tripeptides or dipeptides, LPS	activate caspase and NF-kB transcription factor
	NALPs	Muramyl peptides, bacterial DNA, dsRNA	signal transduction
	RNA helicases	dsRNA	
	Toll-like receptors (TLR3 and TLR9)	dsRNA, vRNA, unmethylated CpG DNA	activate transcription factors
Cell surface	Toll-like receptors	LPS, lipotechnoic acid, flagellin.	activate transcription factor signaling
	mannose receptors	mannose-containing glcyoproteins	phagocytosis
	Fc receptors	antibodies	phagocytosis, degranulation, and signal transduction
	scavenger receptor	negatively charged cell wall components; apoptotic bodies	phagocytosis
	complement receptors	complement components	phagocytosis and chemotaxis

Figure 6.2. Receptors of the innate immune response. The innate immune response can be activated by interaction of specific secreted, cytoplasmic, or cell surface pattern recognition receptors (PRR) expressed on the surface of immune cells with ligands. Ligands are called pathogen-associated molecular patterns (PAMPs) and are expressed or secreted by various pathogens.

A variety of chemical constituents specific to pathogens can activate an innate immune response (Akira et al., 2006). For example, the N-terminal methionine of bacterial proteins is modified by the addition of a formyl group (eukaryotic proteins are not formylated) and N-formylmethionyl peptides derived from bacterial proteins can interact with N-formylmethionyl peptide receptors on a variety of cell types and can activate neutrophils and macrophages. Some bacteria are also able to utilize D-isoform amino acids for some structural components, such as the cell wall, and can be recognized by eukaryotic cells which synthesize only L-amino acids. Double-stranded RNA, derived from replicating viruses, is able to activate a dsRNA-activated kinase that results in the production of an anti-viral response in infected cells (Balachandran et al., 2000). This provides specificity for pathogens since eukaryotic cells do not express long stretches of double-stranded RNA. Unmethylated CpG nucleotides which are present only in bacterial DNA (since in eukaryotes areas of CpG repeats are modified by extensive methylation) interact with monocyte receptors and result in monocyte and macrophage activation.

3. Mechanisms of Recognition by the Innate System – Receptor-Specific Responses

The pathogen-associated molecular structures expressed by microbes are recognized by a group of specific receptors, called pattern recognition receptors (PRRs). PRRs are a diverse group of germ-line encoded protein receptors that are expressed by cells of the innate immune system (Medzhitov and Janeway, 2002: Meylan et al., 2006). PRRs can be classified according to their ligand specificities, function, localization, and/or evolutionary relationships. The activation of PRR leads to the activation/production of the complement cascade, cytokines, and anti-microbial peptide effectors. Activation of PRR on dendritic cells can also promote their ability to present antigen and thereby activate adaptive immunity. Major structural families of PRR include the C-type lectins, leucine-rich proteins, macrophage scavenger protein, plasma pentraxins, lipid transferases, and integrin adhesion molecules (Palsson-McDermott and O'Neill, 2007). On the basis of function, PRRs can be subdivided into signaling PRRs, which include transmembrane Toll-like receptors and cytoplasmic NOD receptors, or endocytotic PRRs that promote the attachment, engulfment, and destruction of pathogens. However, PRRs are most commonly classified by their subcellular localization into three groups: secreted, cytoplasmic, and membrane bound.

I. Secreted PRR

The group of secreted pattern recognition receptors that circulate in the blood, include the pentraxin family proteins (C-reactive protein and serum amyloid), mannose-binding lectin (MBL), complement receptors, collectins, lipid transferase, and peptidoglycan receptors (Garlanda et al., 2005). The C-reactive protein (CRP) is a plasma protein, synthesized by the liver, and induced by inflammation. CRP is able to bind phosphatidyl choline on microbes and can bind complement on foreign or damaged cells to enhance phagocytosis by macrophages that express a receptor for CRP (Pepys and Hirshfield, 2003). The mannose-binding lectin (MBL) can bind a wide range of pathogens via a carbohydrate PAMPs that contains mannose residues, phospholipids, nucleic acids, and some non-glycosylated proteins (Stahl and Ezekowitz, 1998). MBL circulates in blood in a complex with 2 serine proteases, MBL-associated serine proteases (MASP-1 and -2), which bind mannose on pathogens and activate the complement pathway through the mannose lectin pathway (see below) (Takahashi et al., 2008). The interaction of CRP or MBL with PAMPs triggers the activation of the complement system which results in cell lysis following formation of the membrane attack complex, or phagocytosis following release of opsonins.

II. Cytoplasmic PRR

A variety of intracellular proteins can act as receptors for discrete fragments derived from bacterial or viral components. The cytoplasmic Nod-like receptors are a family of approximately 20 proteins, called NODS and NALPS, which have different functions in regulating inflammatory and apoptotic responses (Inohara and Nunez, 2003; Tschopp et al., 2003). These proteins appear to recognize endogenous or microbial molecules (or stress responses) and then form oligomers. The oligomers then activate inflammatory caspases (such as caspase-1) which activate cytokines (such as IL-1) and promote NF-kB signaling. The ligands recognized by these intracellular PRR include bacterial molecules containing muramyl dipeptide and tripeptide moieties, which are found in some bacteria. These ligands bind to the C-terminal leucine-rich repeat (LRR) regions. NOD1 also contains a nucleotide triphosphate-binding domain and a caspase-recruitment domain (CARD) allowing it to interact with signaling molecules, such as RIPK2 (Inohara and Nunez, 2003; Meylan et al., 2006). The result is that binding of muramyl dipeptides and tripeptides to NOD1 promotes cellular signaling and activation of NF-kB. NOD2 is expressed primarily in blood leukocytes and contains similar nucleotide-binding and leucine-rich regions. NOD2 recognizes structures that contain a muramyl dipeptide domain to activate cell signaling and activation of NF-kB. NOD2 can also activate an anti-viral response following exposure to ssRNA or vRNA. The NALPs are a family of 14 proteins that contains a C-terminal leucine-rich region, can bind microbial molecules, and contains a nucleotide-binding region (Tschopp et al., 2003). Although the mechanisms of NALP action are not clear, activators of NALP3 include muramyl dipeptides, bacterial DNA, ATP, toxins, dsRNA, paramyxoviruses, and uric acid crystals and NALP3 mutations underlie some auto-inflammatory diseases. Other cytoplasmic proteins can act as PRR for different pathogen structures. For example, several RNA helicases can act as intracellular PRR (Meylan et al., 2006). The RNA helicases can bind viral dsRNA and ssRNA and then recruit other factors via twin caspase-recruitment domains to activate anti-viral gene programs.

III. Cell Surface Receptors

There are a variety of cell receptors expressed on the plasma membrane of innate immune system cells. These receptors include proteins that: (1) promote endocytosis; (2) act as receptor kinases; and, (3) act as cell signaling receptors (Toll-like receptors). Phagocyte receptors that promote binding and endocytosis of a variety of ligands include; Toll-like receptors (TLR), mannose receptors, complement receptors, Fc receptors, and scavenger receptors that bind a wide variety of polyanion-coated pathogen surfaces (Gordon, 2002; Akira et al., 2006). Mannose receptors, such as those present on the surface of macrophages, can recognize PAMPs which contain mannose-containing glycoproteins. The mannose receptors are cell surface proteins that have large extracellular domains and short cytoplasmic domains that direct receptor endocytosis and recycling (Stahl and Ezekowitz, 1998). These receptors are expressed on macrophages, liver, and endothelial cells and can internalize both pathogen-related glycoconjugates and endogenous glycoproteins secreted from cells as a result of infection or injury (Sahly et al., 2008). When a pathogen is covered with polysaccharides that have terminal mannose residues they bind to these receptors and are engulfed into the phagosome. After attachment to a bacterium, the phagocyte extends "pseudopods" that eventually surround the bacteria and engulfs it into a phagosome. The phagosome fuses with lysosomes which release proteases such as lysozyme, lactoferrin, and cathepsin that destroy the bacteria. During phagocytosis there is an increase in glucose and oxygen consumption which results in a respiratory burst, producing oxygen free radicals capable of killing the pathogens (Greenberg and Grinstein, 2002). Oxygen-dependent myeloperoxidase, NADPH oxidase (produces superoxide), and detoxification enzymes (myeloperoxidase, superoxide dismutase, and catalase) are all enhanced during this process. The production of reactive oxygen species and proteases destroys most bacteria although the tuberculosis and leprosy pathogens can employ the mannose receptor specificity to infect macrophages.

Fc receptors, which bind to circulating antibody-antigen complexes, are found on the surface of macrophages, neutrophils, NK cells, and mast cells (Gordon, 2002). There are several different Fc receptors which recognize antibodies with divergent affinities and cell distributions (Ravetch and Kinet, 1991). The presence of Fc receptors on phagocytes promotes the binding microbes decorated with antibodies resulting in their phagocytosis. Fc receptors expressed on the surface of NK cells can be cross linked by antibodies to promote the release of cytotoxic granules. Similarly, binding of IgE antibodies complexed with environmental antigens to mast cells promotes mast cell degranulation and release of histamine, heparin, and anaphylotoxins (Mekori and Metcalfe, 2002).

Complement receptors refer to a group of integral membrane proteins that bind to proteolytically activated complement proteins to promote phagocytosis or chemotaxis (Rus et al., 2005). For example, the complement protein C3 is activated by proteolytic cleavage into 2 fragments C3a and C3b (see below) and each of these fragments has a specific set of cellular receptors. The C3b fragment binds the surface of pathogens and to several different receptors including complement receptor 1 (CR1), CR2, CR3, and CR4. Thus, C3b binding to pathogens promotes complement receptor-dependent phagocytosis and destruction. CR1 is an integral membrane protein expressed on phagocytes and some lymphocytes (T and B cells) that binds C3b, C4b and iC3b (Caroll, 1998). CR2 is expressed on B cells and dendritic cells (but not on granulocytes or monocytes) and can bind strongly to C3d, iC3b, and weakly to C3b. CR3 and CR4 (also known as β2 integrin adhesion molecules, αMβ2 and αxβ2 respectively) are expressed on granulocytes, phagocytes, and NK cells and binds to iC3b (and ICAM) to promote phagocytosis and adhesion to endothelial cells during leukocyte extravasation.

Toll-Like Receptors

The Toll-like receptors (TLR) are considered central in the activation of immune cells following activation of an innate immune response to exposure to a wide variety of PAMPs (Aderem and Ulevitch, 2000; Ozinsky et al., 2000; Takeda and Akira, 2005). TLR are expressed primarily on macrophages, dendritic cells, granulocytes, lymphocytes (B and T cells), and epithelial cells. In mammals, there are 12 different TLRs that each binds a different class of PAMPs often with the aid of accessory molecules (Figure 6.3). TLRs distinguish the nature of the PAMP and activate specific receptor-mediated pathways to counter the pathogen which frequently involves enhanced cytokine expression and the production of toxic metabolites including oxygen-derived free radicals and nitric oxide (Medzhitov, 2001).

The TLR proteins function as homo- or hetero-dimers and frequently require other protein co-receptors for full ligand specificity. Thus, TLR2 forms a heterodimer with both TLR1 and TLR6 on the plasma membrane to create two distinct cell surface receptors (Ozinsky et al., 2000). The TLR1-TLR2 heterodimer specifically binds the lipoproteins and peptidoglycans that comprise the cell wall of Gram-positive bacteria, such as streptococci and staphylococci. The TLR1 protein is expressed on many cell types and TLR1 expression can be enhanced by exposure to IL-6 and to other cytokines including interferon-α/β (IFN–α/β), IL-10, and TNFα but not by direct exposure to bacterial lipopolysaccharide. TLR2 is expressed on granulocytes and monocytes and is upregulated by exposure to IL-6, IL-10, IFN-α/β, and TNFα as well as by exposure to bacterial lipopolysaccharide. TLR 6 is expressed on monocytes and B cells and can be upregulated by treatment with IFN-γ or IL-

1β. The TLR2-TLR6 heterodimer allows peptidoglycan binding including diacylated lipopeptides, mannose-containing lipoproteins, and lipoteichoic acid (Hajjar et al., 2001). TLR2 also associates with several other proteins including β2 family integrin adhesion molecules (αLβ2, αMβ2) CD14, and possibly some peptidoglycan recognition proteins (Sabroe et al., 2003) to mediate binding to a wide variety of glycolipids, glycoproteins and lipoproteins.

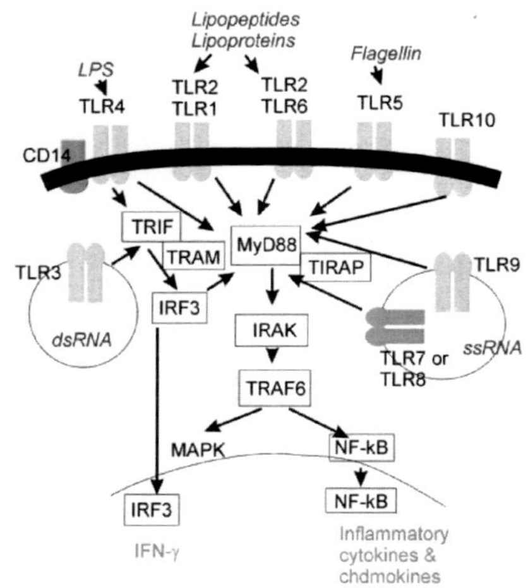

	Ligand	cell type	response
TLR1 + TLR2	bacterial lipoproteins	moncytes, dendritic cells, B cells	IL-6, IL-10, !L-2
TLR2 + TLR6	bacterial lipoproteins, glycolipids, lipoteichoic acid; mannans	monocytes, dendritic cells, mast cells	
TLR2	parasites (?)		
TLR3	dsRNA, viral RNA	monocytes, dendritic cells, B cells	IL-12, IFN-α, IL-6
TLR4 (CD14+LBP)	LPS, heat shock proteins, HMG, viral proteins, fibrinogen, heparin sulfate	monocytes, dendritic cells, mast cells, some epithelia	IL-12, IL-10, IL-6
TLR5	bacterial flagellin	monocytes, dendritic cells, some epithelia	IL-12
TLR7 + TLR8	ssRNA, vRNA	monocytes, dendritic cells, B cells	IL-12, IFNα, IL-6
TLPR9	unmethylated CpG DNA	monocytes, dendritic cells, B cells	IL-12, IL-10, IL-6, INFα

Figure 6.3. Activation of Toll-like receptors. The Toll-like receptors (TLR) are cell membrane receptors expressed on different immune cell types that are central to the activation of the innate immune response following exposure to bacterial antigens. Activation of TLR by ligand binding results in enhanced signal transduction and the activation of the IRF3 and NK-kB transcription factors which promotes expression and secretion of cytokines.

TLR3 is expressed as a homodimer, on dendritic cells or on B, T, or NK cells and its expression can be enhanced by monocyte-derived inflammatory cytokines such as IFN-γ, IL-1β, IL-6, IL-10, and TNFα or by exposure to Gram-positive bacteria. TLR3 appears to be expressed intracellularly, likely on the membranes of the lysosome-endosome compartment. TLR3 binds the dsRNA of viruses engulfed in endosomes to activate signaling (Richer et al., 2009). TLR4 is expressed as a homodimer on a variety of leukocytes, primarily on cells of the monocyte-granulocyte lineage but also on B cells and T cells, usually in association with the monocyte marker CD14. TLR4 expression can be upregulated by exposure to IFN-γ, IL-1β and TNFα but not by direct exposure to Gram negative or Gram positive bacteria. TLR4 indirectly binds bacterial lipopolysaccharide (LPS) derived from the cell walls of Gram-negative bacteria (Alexander and Rietschel, 2001). LPS binds to the LPS-binding protein (LBP) which directs the LPS to CD14, associated with the membrane, and results in activation of associated TLR4. Additionally, TLR4 interacts with cell type-specific molecules including the chemokine receptor CXCR4, GDF5, CD55, heat shock proteins, and some complement receptors to promote efficient LPS-dependent signaling. The TLR4 complex also recognizes lipoteichoic acid, several viruses, including respiratory syncytial virus, hepatitis C virus, and murine mammary tumor virus, and endogenous ligands including heat shock proteins, fibronectin, fibrinogen, surfactant protein A, and β-defensins (Miller et al., 2005). Further, there is some evidence to suggest that a heterodimer of TLR4 and TLR1 can inhibit

TLR4 activity while a TLR5-TLR4 heterodimer enhances signaling (Nishiya and DeFranco, 2004).

TLR5 exists as a homodimer (or as a heterodimer with TLR4) to bind to flagellin, the major protein composing the flagella of motile bacteria (Hayashi et al., 2001). TLR5 is expressed at the highest levels on monocytes, but is expressed on other leukocytes, and can be enhanced by exposure to IL-6, IL-10, and TNFα, and by exposure to Gram-positive and Gram-negative bacteria. TLR7 and TLR8 form a heterodimer that binds the ssRNA genomes from viruses including influenza, measles, and mumps engulfed in endosomes. TLR7 is expressed primarily on monocytes, dendritic cells, and B cells while TLR8 is highest in monocytes and their expression can be enhanced by treatment with IL-6 or the monocyte-derived cytokines (Gorden et al., 2006). The TLR9 homodimer is expressed both intracellularly and on the plasma membrane of B cells, dendritic cells, or monocytes. TLR9 binds to the unmethylated CpG DNA of bacteria engulfed in endosomes (Ahmad-Nejad et al., 2002). TLR10 and TLR12 are expressed on B cells and monocytes and are enhanced by treatment with multiple cytokines but have an unknown ligand-binding specificity. TLR11 appears not to be expressed on human cells; it can mediate murine cell binding to protozoan proteins.

Binding of the TLR to their ligands promotes activation of an overlapping set of signaling molecules that results in activation of the NF-kB transcription factor and promotes secretion of TNFα, IL-1β, and chemokines that all result in the promotion of a localized inflammatory response (Aderem and Ulevitch, 2000; Medzhitov and Biron, 2003). The TLR receptors interact with one or more of 4 possible adapter proteins, including MyD88, TIRAP, TIRF, and TRAM, to promote signal transduction (Yamamoto et al., 2002). All of the TLR receptors, except TLR3, signal through MyD88 (Jiang et al., 2003). The TLR4 and TLR2-containing heterodimers also activate TIRAP. TLR3 signals through TRIF. The adapter proteins then interact with and activate specific protein kinases within the cell, including IRAK, IRAK4, TBK1, and IKKi. Activation of these kinases leads to amplification of the signal resulting in activation of transcription factors and changes in gene transcription for specific target genes. The result is that the large diversity of TLR ligands, which include nucleic acids, proteins, lipids, and polysaccharides, derived from pathogens can all activate the expression of multiple cytokines, effector proteins, and free radical generation systems to promote a potent and generalized immune response.

4. Cell Types and Processes Involved in the Innate Immune System

I. Innate Barriers to Infection

There are a variety of cell types involved in the innate immune system. Since the role of the innate immune system is to prevent infection or rapidly resolve infections by pathogens, these cells can provide a physical barrier against pathogens, or can actively recognize, destroy and clear pathogens. Epithelial cell layers are the most obvious components of the innate immune system that can provide a physical and chemical barrier against pathogen infection

(Figure 6.4). The physical barrier formed by epithelial surfaces is normally impermeable to most infectious agents. For example, the skin prevents pathogens from invading the organism and the normal turnover of skin epithelium (desquamation) improves this efficiency by removing agents which have adhered (Elias, 2007). The epithelial cell layers involved in maintaining a barrier against pathogen infection secrete mucous which is constantly being shed by the activity of cilia. The mucous secreted by the mucosa-associated lymphoid tissues (MALT), that include the epithelial cell layers of the respiratory, digestive, and urogenital tracts, the eye conjunctiva, and the ducts of all exocrine organs, protects against invasive pathogens or particles in the air, diet or environment (Acheson and Luccioli, 2004). In a healthy adult the MALT contains approximately 80% of all immune cells and constitutes the largest lymphoid organ system. The trapping effect of the mucous lining produced by the epithelium of the lung and gastrointestinal tracts coupled with peristalsis helps keep the air passages and gastrointestinal tract free from pathogens. Further, the production of tears and saliva promotes clearance of pathogens by flushing.

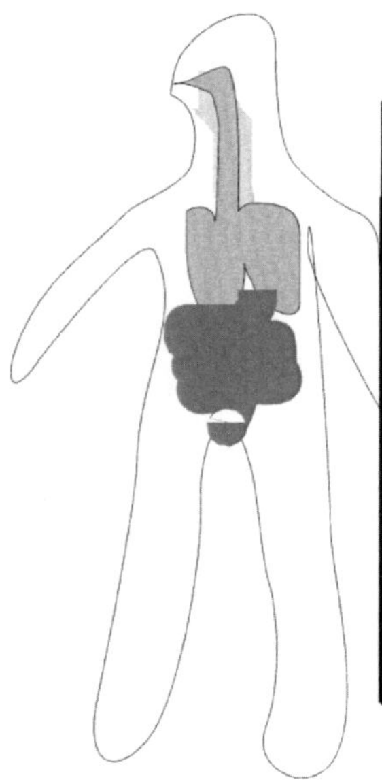

	Skin	Gut	Lungs	Eyes & Nose
Mechanical	Epithelial cells joined by tight junctions			
	Desquamation	Longitudinal flow of air or fluid Movement of mucous by cilia		Flushing with tears & mucous
Chemical	Fatty acids	Low pH		Salivary enzymes (lysoszyme)
		Enzymes (pepsin)		
	Antibacterial peptides (defensins)			
Microbiological	Commensal flora			

Figure 6.4. Physical barriers to infection are important components of the innate immune system. Organ surfaces that are susceptible to pathogen infection, such as the skin, gut, lungs, eyes and nose, possess mechanical, chemical, and microbiological barriers to inhibit pathogen colonization.

These mucosal sites also have strong chemical barriers and cleansing function to repel pathogens. Fatty acids present in sweat can inhibit the growth of bacteria and contribute to the chemical barrier function of epithelial cell layers. Epithelial cells in the lung and gastrointestinal tracts can also produce a variety of chemical agents which have anti-microbial activity. For example, following injury or in response to inflammatory cytokines, epithelial cells produce defensins (a group of small cysteine-rich peptides) and cathelicidins which have anti-microbial activity (Ganz, 2003; Zanetti, 2004). Defensins are secreted and processed from a larger precursor by epithelial cells, neutrophils and some leukocytes and are found in skin, and the lining of the gastrointestinal tract, genitourinary tract, and nasal passages and lungs. The defensin molecules are able to create holes in the plasma membrane and kill pathogens (Ganz, 2003). Cathelicidins, such as LL37, are synthesized by macrophages, neutrophils, and epithelial cells and provide anti-microbial protection to the skin and lining of the urinary tract (Zanetti, 2004). Further, lysozyme, which breaks β(1,4) glycosidic linkages in peptidoglycans, and phospholipase which can break down phospholipids, are present in tears and saliva and can disrupt bacterial cell walls and membranes (Dajani et al., 2005).

It has become clear that normal flora, provided by commensal bacteria, contributes to the innate immune barrier function. For example, the normal commensal bacteria in the gut induce anti-inflammatory signals in the epithelium-associated mucosa that protects the gut from pathogens (Tlaskalova-Hogenova et al., 2004). This response appears to be mediated by alterations in inflammatory T helper cell (T_H1) responses. When the gut is depleted of normal flora, an abnormal loss of T_H1 function results but when the normal flora is restored, normal T_H1 function is restored. The intact gut barrier provides an immune response to eliminate pathogens; however, a compromised gut barrier permits immune responses to commensal flora resulting in cytokine production, T_H1 activation, and inflammatory bowel diseases such as Crohn's disease or immune responses to food allergens such as gluten to cause celiac disease.

II. Innate Immune Cells

Cells of the innate immune system, such as white blood cells including neutrophils and monocytes, can also actively destroy and clear pathogens. The epithelial cell layers are also populated by a variety of cell types including specialized macrophages, intraepithelial T and B cells and mast cells which are able to recognize microbes and recruit a larger response to infection. Exposure to pathogens results in the production and release of various factors which attract other leukocytes to the site of infection (Lin and Karin, 2007). Leukocytes, or white blood cells, are circulating cells that move freely through blood vessels and can interact with and capture cellular debris, foreign objects, or invading pathogens. Leukocytes arise from the differentiation of pluripotent cells in the bone marrow and are unable to replicate in the vasculature. The principle leukocytes of the innate immune system include phagocytes such as neutrophils and monocytes and cells which produce anti-microbial enzymes such as natural killer cells, basophils, and mast cells (Figure 6.5).

Phagocytes, the most numerous of the cells of the innate immune system includes neutrophils and monoycte/macrophages (Dale et al., 2006). The normal role of phagocytes is to phagocytose pathogens or particles during tissue development and maintenance. They are also able to remove cells which have undergone apoptotic cell death, without activating an inflammatory or adaptive immune response. The apoptotic bodies have a negative surface charge and are targeted for phagocytosis by macrophages via "scavenging receptors" (Hart et al., 2004). Similarly, pathogens are recognized by the phagocyte and are enveloped in an out-folding of the plasma membrane to generate a phagosome. The ability of a phagocyte to recognize a pathogen is enhanced by the presence of "opsonins" which include several complement components and antibodies. The opsonin binds to both the pathogen and to specific receptors on the phagocyte including complement receptor and Fc receptors (Gasque, 2004; Hart et al., 2004). Phagocytes respond to a variety of cytokines and chemokines which are able to promote phagocyte activation and accumulation to sites of damage.

	cell type	**Properties**
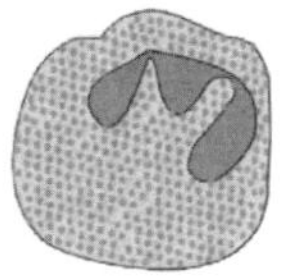	**Neutrophil**	• phagocytose and kill bacteria • produce anti-microbial peptides • degranulate to release histamine, heparin and porins
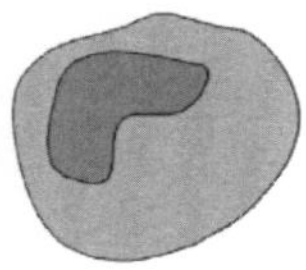	**Monocytes**	• phagocytose and kill bacteria • produce pro-inflammatory cytokines • release anti-microbial peptides
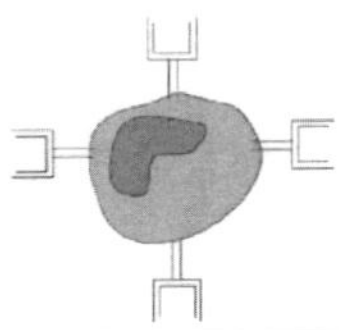	**Dendritic cells**	• antigen presentation • secrete cytokines (IFNα, IL-10, IL-12)
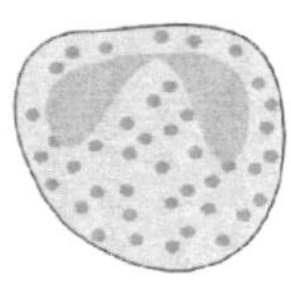	**Eosinophil**	• kill invading parasites • secrete heparinases
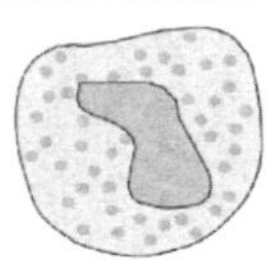	**Basophils and Mast cells**	• degranulate to release heparin, histamine, bradykinin, prostaglandins • release cytokines (TNFα, IL-6, IFN-γ)
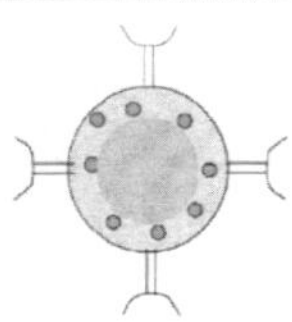	**Natural killer cells**	• kill foreign and host cells with low surface MHC • release cytokines (TNFα, IFN-γ)
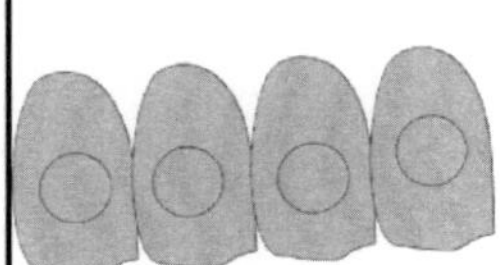	**Epithelial cells**	• produce anti microbial peptides • tissue-specific mediators (mucous, tears, sweat, etc)

Figure 6.5. Cells of the innate immune response. Several distinct cell types are involved in the activity of the innate immune system. Neutrophils are the most common cells and act to destroy and phagocytose pathogens. Monocytes and dendritic cells are phagocytes which can present antigen and promote adaptive immunity while granulocytes, such as eosinophils and basophils, promote immune cell influx, and natural killer cells kill infected host cells.

Neutrophils are the most abundant population of circulating leukocytes comprising 50-70% of the leukocytes in the blood. A non-activated neutrophil has a half-life of less than 12 h in blood (approximately 10^{11} neutrophils are produced/day). Neutrophils mediate the earliest phases of the innate immune response and are characteristic of an inflammatory response (Burg and Pillinger, 2001). Chemotactic agents released at the site of infection or trauma (such as cytokines including IL-8 and IFN-γ or activated complement components such as C3a) promotes the activation and accumulation of neutrophils within minutes followed by their migration into the underlying tissues where they can survive for a few days. Upon activation, neutrophils are induced to phagocytose pathogens or particles or to "degranulate" releasing granules filled with proteolytic enzymes (Kobayashi et al., 2005). Neutrophils consume multiple microbes, each creating a phagosome where hydrolytic enzymes and oxygen-derived free radicals destroy the invading organism. The production of free radicals, called the "respiratory burst", involves a variety of enzyme systems, including NAPDH oxidases, superoxide dismutases, and myeloperoxidases, to change oxygen to bactericidal oxygen radicals and peroxides. The process of degranulation involves the release of preformed cytoplasmic granules into the extracellular space. There are three types of granules each with a different subset of active materials: specific granules contain lactoferrin, lysozyme, and cathelicidin; azurophilic granulocytes contain myeloperoxidase, lysozyme, elastase, cathepsin, and defensins; and, tertiary granules contain cathepsin and gelatinases. In addition, neutrophils secrete a complex mixture of granule-associated proteins and DNA which creates a neutrophil extracellular trap (NET) that physically entraps bacteria and promotes their destruction (Brinkmann et al., 2004). Thus, the role of neutrophils in an innate response is to provide a rapid and strong response to infectious agents that includes the release of agents that can directly kill pathogens and to recruit other immune cells.

Monocyte/macrophages are phagocytotic leukocytes which are less numerous than neutrophils but that can survive longer. While circulating in blood, these cells are most commonly referred to as monocytes and can survey outside of the vascular by moving across capillary vessels. However, when they are present in tissues, they can differentiate into organ-specific macrophages. In circulation monocytes have a half-life of approximately 1-3 days while once they differentiate into tissue macrophages they have a half-life of many years. Virtually every tissue has a resident population of tissue macrophages in the subepithelial connective tissue, in the interstitia of the parenchymal organs, in the lining of the vascular sinusoids of the liver and spleen; eg. Kupffer cells, alveolar macrophages, sertoli cells, astrocytes, etc. (Laskin et al., 2001). Macrophages are recruited early and maintained at the site of inflammation. Monocytes/macrophages are a major source of cytokines and are the important effector cells in the innate response (and adaptive responses) (Serbina et al., 2008). For example, macrophages produce IL-1β and TNFα to promote monocyte activation, secretion of matrix-metalloproteinases, and upregulation of adhesion molecules and chemokines to activate and attract other immune cells. Macrophages can also produce IL-12 which promotes T_H1 recruitment and activation. Like neutrophils, macrophages are able to phagocytose substantial numbers of pathogens and are able to destroy the phagocytosed bacteria by production of hydrolytic enzymes and oxygen-derived free radicals (Aderem and Underhill, 1999). In addition, macrophages release a variety of inflammatory mediators including prostaglandins, leukotrienes, cytokines and chemokines which attract additional effector cells but, unlike neutrophils, macrophages do not contain and release granules of active proteolytic enzymes. Monocyte/macrophages are the first line of anti-pathogen defense.

Dendritic cells are specialized tissue phagocytes present at low levels in the skin (where they are called Langerhans cells), and inner mucosal lining of the nose, lungs, stomach, and intestines (Serbina et al., 2008). Their role in the skin and epithelium is to act as a surveillance mechanism for tissue damage and pathogen infection. Dendritic cells are derived from hematopoietic precursors and may have either myeloid (macrophage) or lymphoid character. Dendritic cells are key initiators of both innate immune responses, via cytokine production, and of adaptive immune responses, via presentation of antigen to T cells (Banchereau and Steinman, 1998). The immature dendritic cell is highly phagocytotic and continuously samples the environment. Once it encounters a pathogen, high levels of cytokines are produced that can activate macrophages and recruit other phagocytes and effector cells to the site of injury as part of the innate system. In addition, the phagocytosis of a "presentable antigen" promotes activation of the dendritic cell, digestion and representation of fragments of the antigen on the dendritic cell surface, and its migration to lymphoid tissues (such as lymph nodes) where it can present the antigen to activate T_H1 cells and initiate an adaptive immune response.

Natural killer cells are large granular lymphocytes that can directly kill target cells including virally-infected cells or tumor cells, which express low levels of human leukocyte antigen (HLA). The natural cytotoxicity receptors (NCR) expressed on NK cells mediate activation against target cells by activating networks of signaling pathways. NK cells directly kill virus-infected cells without additional stimuli following recognition of antibody-coated cells, virus infected cells, or cells that do not express MHC class I dependent-inhibitory pathways (Biron et al., 1999); NK cells express receptors that bind HLA and then inhibit NK function. Natural killer cells attack host cells that have been infected with pathogens but do not directly attack pathogens and are not phagocytic (Cooper et al., 2001). NK cells are activated by cytokines, such as IL-2, IFN-γ, IL-1β, and IL-6 produced by virally-infected cells. Activation of Fc receptors results in NK degranulation while expression of killer immunoglobulin-like receptors (KIR) bind human leukocyte antigen on the target cell to inhibit degranulation. NK cells contain granules rich in perforin which, when released, can form pores in nearby cells inducing lysis and granzymes which enter the cell and promote apoptosis or granulolysin which destroys intracellular bacteria.

Granulocytes are important amplifiers of innate immune responses and are present in nearly all inflammatory responses. Granulocytes are derived from bone marrow stem cells and differentiate into neutrophils, eosinophils and basophils and unchecked accumulation and activation of granulocytes can lead to tissue damage. Eosinophils express Fc receptors for antibody and can act as potent cytotoxic effector cells for various parasitic organisms, such as helminthes. Degranulation of eosinophils results in the release of major basic protein, eosinophil cationic protein, and eosinophil-derived neurotoxin which can directly damage the parasites (Jacobsen et al., 2007). Interestingly, eosinophil granules also contain anti-inflammatory enzymes, including histaminases and phospholipase, to down-regulate or terminate an ongoing inflammatory response. In contrast, basophils are able to release histamine, leukotrienes, chemotactic factor of anaphylaxis, and neutral protease, to promote an allergic hypersensitivity response following IgE binding to high affinity Fc receptors (Wedermeyer et al., 2000). Basophils can also be activated by binding to a pathogen and are capable of producing anti-bacterial free radicals.

Mast cells reside in connective tissues and mucous membranes. While usually associated with allergy and anaphylaxis, mast cells have potent anti-microbial activity and are important in wound healing and repair (Mekori and Metcalfe, 2000). Upon activation, mast cells rapidly release granules which contain high levels of histamine, heparin, hormone modulators, and chemokines similar to basophils (Wedermeyer et al., 2000). This results in the dilation of blood vessels and recruitment of neutrophils and monocytes which promotes the characteristic signs of inflammation (swelling, redness, and heat). Mast cells also release inflammatory cytokines, including TNF-α, IL-6, and IFN-γ, in response to pathogen which recruit and activate monocytes and granulocytes.

III. Complement System

The Complement system of cascading plasma enzymes and effector proteins includes more than 20 proteins that are produced primarily in liver hepatocytes and which account for about 5% of the globulin fraction of serum. These proteins work together as a biochemical proteolytic cascade where the activation of one component promotes its ability to proteolytically cleave and thereby activate the next component in the cascade (Nauta et al., 2004; Rus et al., 2005). The complement system acts to "complement", or help, the ability of antibodies to clear pathogens (Carroll, 2004). Activation of the entire cascade results in the production of the "membrane attack complex, a pore in targeted cells to promote cellular lysis. However, proteolytic cleavage of various components also produces a number of active proteins that are important effectors in the inflammatory response. For example, some of these components act as chemokines to trigger the recruitment of inflammatory cells. Other cleaved components act as opsonins and bind to foreign bodies (or cells) and target them for phagocytosis and destruction by neutrophils and monocytes.

The complement cascade can be initiated by the presence of antibodies or pathogens (Gasque, 2004). The three general means of activating the complement system are the classical pathway, the mannose-binding lectin pathway (MBL-MASP), and the alternative pathway (Figure 6.6). The classical pathway of complement activation depends on the binding of antigen-antibody complexes (IgM > IgG) to the complement C1 complex resulting in the activation of C1 (Carroll, 2004). The C1 complex is made up of C1q, two molecules of C1r, and two molecules of C1s. The antigen-antibody complex binds and aggregates C1q which activates the serine protease activity of C1r and cleaves and activates the serine protease activity of C1s. The activated C1 complex cleaves complement components C2 and C4 to generate C2a, C2b, C4a, and C4b. C4b and C2b combine to form an active complex (called the C3 convertase) that proteolytically cleaves complement component C3 to generate C3a and C3b. C3b forms an active complex with C4bC2b to create the C5 convertase complex that cleaves complement component C5. C3a (and C5a) act as leukocyte chemotactic proteins that promote activation and enhancement of an inflammatory response. The mannose-binding lectin pathway is very similar to the classical pathway of complement activation. The lectin pathway involves binding of the complex comprised of the mannose-binding lectin (MBL) and MBL-associated serine protease (MASP) -1 and-2 to mannose residues on the surface of the pathogens (Fujita et al., 2004). (The MBL and MASP proteins are similar in structure and function to the components of the C1 complex). This binding activates the MASP protease activity which results in the cleavage of C4 and C2 to generate C4a, C4b, C2a, and C2a and C4b then binds to the bacteria and activates the C3 convertase which eventually creates the membrane attack complex to lyse the bacteria.

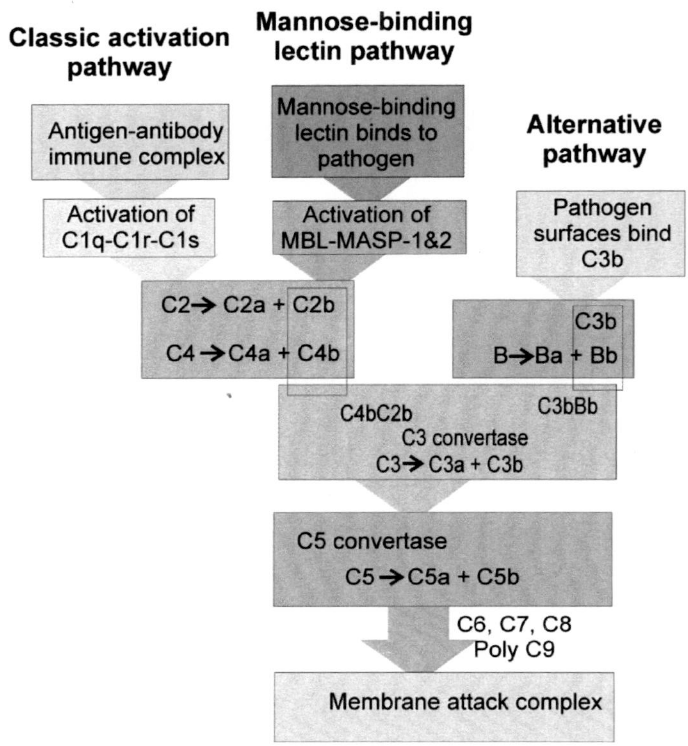

Figure 6.6. Activation of the complement system. The complement system is a series of plasma proteins and effector proteins that can be activated in response to three distinct pathways: classical, mannose binding lectin, and alternate pathways. Activation of the complement pathway creates active proteins that promote immune cell accumulation, enhance phagocytosis, and create the membrane attack complex to lyse target cells.

The alternate pathway of complement activation depends on the spontaneous breakdown of complement component C3 which occurs at a slow but constant rate such that low levels of C3a and C3b are present in circulation. The C3b complex is able to bind any pathogen in the environment which changes its conformation and promotes binding to factor B. In the absence of pathogen, C3b reassociates with C3a. The presence of factor D cleaves C3b-bound factor B to create Ba and Bb to generate the C3bBb complex which acts as the "alternate C3 convertase" to cleave C3 to C3a and C3b. In the alternate pathway, C3bBbC3b acts as the C5 convertase to activate the remainder of the pathway. The three pathways of complement activation all converge on the terminal pathway. C5b initiates the activation of C6, C7, C8, and the multimeric C9 protein that result in the production of the membrane attack complex (MAC). C5b binds to the cell membrane and interacts with C6 to form a stable complex. C5bC6 binds C7 and the hydrophobic core inserts the C5bC6C7 complex into the cell membrane where it binds C8. The C5bC5C7C8 has weak cell lysis activity but is able to bind up to 15 subunits of C9 to form the MAC (Gasque, 2004). Thus, this complex is inserted into the membranes of the target cell creating physical holes that result in cell lysis.

IV. Cytokine and Chemokine Production

A wide variety of cytokines and chemokines are produced by monocytes, neutrophils, NK cells (and endothelial cells, keratinocytes, hepatocytes, etc) which are involved in regulating immune responses (Lin and Karin, 2007). Cytokines are small secreted proteins which mediate and regulate immunity, inflammation, and hematopoiesis. Chemokines are a group of cytokines which promote directed migration (chemotaxis) by responsive cells – chemotactic cytokines. Cytokines are produced by specific cells *de novo* in response to immune stimuli and usually act over a short distance either on the same cells that produced them (autocrine) or on a different set of responsive cells in the same area (paracrine). Cytokines act by binding to specific membrane receptors on the surface of responsive cells which promotes the activation of a signaling cascade which can differ based on the receptor (although frequently involving activation of tyrosine kinases) which can alter cell behavior and the expression of genes including those for membrane receptors, cytokine receptors, cell proliferation elements, and various secreted effector molecules (Hawiger, 2007). In general, cytokines mediate communication among leukocytes and can act as modulators following recognition of specific elements. Frequently, cytokines amplify responses following recognition of specific elements to promote the activity of a wider variety of cell types (Mizgerd et al., 2001).

The role played by cytokines can be quite complicated. Some cytokines can be secreted by a variety of different cell types: IL-6 can be produced by macrophages, lymphocytes, and stromal cells. A single cytokine can bind to and act on multiple cell targets sometimes with different effects: for example, IL-1 and IL-6 can promote the activation and differentiation of T and B lymphocytes, activate adhesion molecules on endothelial cells, and promote acute phase responses by stromal cells. Cytokine effects can be redundant meaning similar functions can be activated by distinct cytokines: both IL-1 and TNFα can promote monocyte activation and extravasation (Mizgerd et al., 2001). Cytokines are also frequently involved in cascades where treatment with one cytokine activates the target to produce additional cytokines. It is also well known that in cells exposed to multiple cytokines, the cytokines can act synergistically or antagonistically sometimes generating positive or negative feedback loops. A variety of inflammatory cytokines, including IL-1, IL-6, IL-8, IL-12, and TNFα are produced by macrophages, dendritic cells, and some stromal cells (Heller et al., 1997); IL-1, IL-6 and TNFα promote acute phase responses and activate adhesion molecule expression and protease secretion by leukocytes and endothelial cells, while IL-8 is a chemoattractant for neutrophils and monocytes and IL-12 is an activator of the adaptive immune response.

5. Inflammatory Responses

While innate immunity is defined as the rapid response initiated by activation of a small group of genetically encoded receptors, these processes are also able to active components of the adaptive immune response to enhance the clearance of infecting organisms. One step in the escalation of an immune response to an infecting pathogen is the inflammatory response. When the physical barrier of the epithelial cell layer is breached by a pathogen, it immediately encounters the phagocytes and granulocytes present in the tissues. Specific pathogen components (PAMPs) are recognized by a small group of genomically encoded receptors (PRR) on the phagocytes, which promotes phagocytosis and the production of cytokines and chemokines, or on granulocytes, which promotes their degranulation resulting in the release of vasodilators and anti-coagulants (histamine and heparin), cell lysis factors (perforin), and other immune activators (prostaglandins, bradykinins, and cytokines) (Medzhitov and Biron, 2003; Akira et al., 2006; Lin and Karin, 2007). The continuing presence of pathogens results in an amplification of the immune response which results in an inflammatory response (while clearance of the pathogen allows the response to resolve).

Inflammation refers to a complex process after pathogen infection and recognition by phagocytes or granulocytes (macrophages, dendritic cells, neutrophils, and mast cells) that result in the recruitment of leukocytes to sites of infection and the production and release of a wide range of bioactive compounds that alter local blood flow and alter tissue remodeling (Figure 6.7).

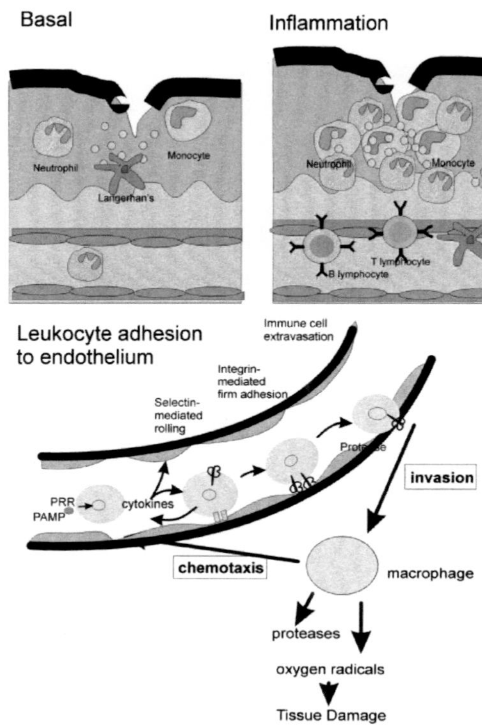

Figure 6.7. Diagrammatic summary of an inflammatory response. Activation of surveilling Langerhan's cells, monocytes, and neutrophils in response to bacterial antigens results in increased vascular permeability and accumulation of immune cells including neutrophils, monocytes, and lymphocytes to eliminate the infectious agent. The accumulation of leukocytes to extravascular sites of infection involves adhesion via specific selectin or integrin family adhesion molecules to the blood vessel wall, followed by cell invasion into tissues where they produce oxygen radicals and proteases to destroy infecting pathogens.

Inflammatory responses are generally associated with the five diagnostic signs: rubor (redness); calor (increased heat); tumor (swelling); dolor (pain); and functio laesa (loss of function) (Nathan, 2002). The increased redness and heat in the area of inflammation are associated with an increase in blood flow to the inflamed site in response to the release of vasodilators and anticoagulants. The increase in local blood flow and the increase in vascular permeability allow plasma to move into extravascular tissue which causes the local swelling. Pain is due to the release of chemicals (such as bradykinin) that sensitize nerve endings: however, acute inflammation of the lung is not usually associated with pain because there are no appropriate sensory nerve areas in lung tissue. Acute inflammation refers to the initial response of the body to pathogen infection while chronic inflammation refers to a prolonged response that results in a shift in the type of cells present and is characterized by simultaneous destruction and healing of the tissue which can often become pathogenic. In general, inflammation is an early response by the immune system to infection or irritation. Injured cells or pathogens release several factors, including cytokines, chemokines, and vascular dilators, which promote the accumulation of leukocytes to the site of infection (Nathan, 2002). In addition to cell-derived processes, several biochemical cascades are also activated in an inflammatory response. The complement system is activated by pathogen, while the coagulation and fibrinolysis systems are activated by necrosis.

A wide variety of cytokines are produced by monocytes, neutrophils, NK cells (and endothelial cells, keratinocytes, hepatocytes, etc) to recruit and activate leukocytes and produce systematic alterations including increases in the synthesis of effector cells and proteins that potentiate antimicrobial and anti-viral responses (Hawiger, 2007). Bacterial infection, viral infection, and tissue damage all lead to cytokine production, principally the pro-inflammatory cytokines IL-1β, IL-6, and TNF-α by monocytes and endothelial cells in the vicinity of the damage (Mizgerd et al., 2001). Exposure to these cytokines promotes the expression of proinflammatory genes including adhesion molecules, proteases, and growth stimulatory activities and the activation of monocytes and other lymphocytes (Hawiger, 2007). These cytokines also activate endothelial cells which are active collaborators in the inflammatory process - one of the major effects of cytokine treatment is the production of multiple leukocyte adhesion molecules on the surface of endothelial monolayers. Leukocyte/endothelial cell adhesion is important to localizing active leukocytes to areas in the proximity of damage. Chemokines are also produced and released from injured cells or a wide variety of cells exposed to bacterial LPS or cytokines. Although many chemokines are released into plasma they can bind heparin sulphate proteoglycans on the surface of endothelial cells which keeps them near the site of synthesis. Exposure to chemokines promotes leukocyte migration to areas of higher chemokine concentration and increases the affinity of integrin adhesion molecules to stabilize leukocyte interactions with the endothelium. This results in the accumulation and activation of leukocytes into the proximity of bacterial infection, viral infection, or cell damage which then produce cytotoxic factors that fight infections.

Leukocyte adhesion to endothelial monolayers involves a variety of receptor/ligand pair interactions which may differ depending on leukocyte or activation. Treatment of endothelial cell monolayers with inflammatory mediators promotes the activation and expression of several selectin and immunoglobulin adhesion molecules. For example, P-selectin is upregulated within minutes (preformed receptor is cycled to the membrane) and E-selectin is upregulated at the transcriptional and translational levels and expression is significantly enhanced within 2 h (Nathan, 2002; Hawiger, 2007). Selectins are particularly important as adhesion molecules which utilize their low affinity high rate adhesions to slow leukocytes down as they pass through the blood vessel. E-selectin mediates binding to neutrophils and monocytes and selectin ligands on leukocytes are principally glycoprotein side chains such as Lex and Ley blood antigens. Endothelial immunoglobulin family adhesion molecules including ICAM-1 and VCAM-1 are upregulated at the transcriptional and translational levels and expression is significantly enhanced by treatment with inflammatory stimuli. ICAM-1 is upregulated within 12 h and is maximally expressed at 24 h and is the major mediator of monocyte adhesion, but also mediates adhesion to other leukocytes which express the β2 integrin adhesion molecule ligands. ICAM-1/β2 integrin interactions are high affinity interactions which stop leukocyte flow and promote leukocyte motility (signal transduction) and extravasation. VCAM-1 expression is upregulated with 24 h, is maximal at 36 - 48 h, and is maintained for several days and acts as the major adhesion molecule of T cells (which accumulate at the site of inflammation later in the process). The VCAM-1 ligands on the T cells are the α4β1 and α4β7 integrins which mediate high affinity interactions and which promote T cell extravasation. Leukocyte extravasation is the combination of leukocyte activation and increased motility and proteolytic destruction of the vascular extracellular matrix in response to cytokine activation and signaling resulting from leukocyte/endothelial adhesion. Thus, the result of the inflammatory process is the accumulation of activated immune cells (phagocytes, granulocytes, and lymphocytes) into the area of infection. These cells resolve the infection by phagocytosis of the microbes and the production of various toxic metabolites which destroy the pathogens.

6. Antigen Processing - Role of Innate Immunity in Stimulating Adaptive Immune Responses

The innate response provides signals which are important in stimulating the adaptive immune response. In the two-step hypothesis of adaptive immune activation: step 1 is antigen presentation to the lymphocytes by activated phagocytes, and step 2 is a co-stimulatory activation of the target lymphocyte by a second receptor system or cytokine. Macrophages and dendritic cells are the principal phagocytes that "present" antigens to T cells to initiate both cell-mediated and humoral (antibody-mediated) adaptive immune responses (Harding and Song, 1994). As part of their normal function in the innate immune system, these antigen-presenting cells recognize and phagocytose the pathogen and destroy and digest it in the lysosomal system. The digested fragments of the engulfed pathogens can be returned to the phagocyte surface in association with the antigen-presenting cell (APC) major histocompatibility complex (MHC) antigen. The macrophages and dendritic cells (Langerhans cells) then migrate to lymph nodes where they encounter lymphocytes of the adaptive immune system. The antigen-MHC complex on the APC then binds to the T cell receptor on a receptive T cell (which recognized the specific pathogen fragment – antigen) to initiate T cell activation. Complete activation requires the presence of costimulators which mediate interactions between the APC and T cells. Costimulators, such as the B-7 (CD80) and B7-2 (CD86) membrane proteins, are induced on the surface of APCs in response to PRR/PAMP interactions. Activation of dendritic cells by interaction of PAMPs with TLR promotes the secretion of cytokines including IL-12 (T_H1), IL-23 (T_H17), and IL-6. B cells also express TLRs and are able to bind PAMPs and act as APCs. Further, pathogens coated with fragments of complement C3 are not only opsonized for phagocytosis but bind more strongly to B cells that have bound pathogen through their BCR and thus promote B cell activation.

7. Disorders of the Innate Immune System

The importance of the innate immune system in protecting individuals from pathogens is evident from the number of disorders and diseases that result from dysfunction in the various elements. Mutations in the genes that encode proteins important in an innate response such as complement, cytokines, enzymes, and receptors can increase susceptibility to infection. In general these disorders result from mutations that cause a dysfunction in the ability of immune cells to clear infecting organisms. For example, congenital disorders that affect phagocyte function include those characterized by problems with phagocyte movement, phagocytosis, or phagocyte-dependent killing of microbes (Lekstrom-Himes and Gallin, 2000). Mutations in the phagocyte cytoskeleton that alter lysosomal trafficking are associated with Chediak-Higashi syndrome while mutations in actin polymerization are associated with Lazy Leukocyte Syndrome. Mutations in phagocyte adhesion molecules, for example in the β2 integrin, result in Leukocyte Adhesion Deficiency. Deficits in free radical production resulting from mutations in NADPH oxidase (Chronic granulomatous disease), glucose-6-phosphate dehydrogenase, or myeloperoxidase all result in decreased pathogen killing. Job's Syndrome is the result of a deficiency in IFN-γ production and causes defective phagocyte chemotaxis. The result of all of these congenital disorders in phagocyte function is that affected individuals show a significant incidence of infection with a wide variety of organisms including pyogenic bacteria, *S. aureus*, *S epidermidis*, yeasts, and fungus. Langerhan's cell histocytosis (LCH) is a disorder marked by excessive numbers of Langerhan's cells and results in damage to skin, bone and other organs. Alterations in dendritic cell function are also involved in allergy and autoimmune disorders like systemic lupus erythematosus (SLE) and inflammatory bowel diseases.

A variety of congenital disorders are associated with a decrease in neutrophil number (neutropenia) (Burg and Pillinger, 2001; Berliner et al., 2004). Since neutrophils are the first innate immune cells to encounter a pathogen, neutropenia can promote significant life-threatening symptoms. Mutations in neutrophil elastases (proteases that cleave connective tissue proteins and allow migration into underlying tissue) such as Kostman syndrome or Shwachman-Diamond syndrome promote neutropenia and increase mortality by 15-25% while cyclic neutropenia which shows variable effects on neutrophil number can be life-threatening in 10% of subjects. Myelokathexis is associated with a defect in the chemokine receptor, CXCR4 which promotes neutropenia and abnormal neutrophil morphology (Berliner et al., 2004). Although it is unknown what gene mutations are involved, the destruction of neutrophils is associated with chronic autoimmune neutropenia and chronic idiopathic neutropenia (postulated to be the result of an autoimmune anti-neutrophil antibody). Changes in neutrophil number also result from a variety of environmental causes or as side effects of other diseases in otherwise normal individuals. For example, infection with some pathogens or autoimmune disorders can consume neutrophils creating neutropenia. Leukocytic neoplasms frequently result in neutropenia while the chemotherapy agents used in the treatment of neoplasms almost always destroy significant neutrophil numbers. The result of neutropenia from any cause is an increase in bacterial infections which are frequently life-threatening.

A variety of mutations in the pattern recognition receptors that allow the immune system to recognize pathogens have been shown to be associated with clinical consequences. These mutations allow some pathogens to go unrecognized by the immune system or can result in a hyper-responsiveness which can generate autoimmune disease. Interestingly, mutations in NOD2 have been linked to inflammatory bowel disease and Crohn's disease (Inohara and Nunez 2003). Mutations in NALP3 underlie auto-inflammatory diseases such as cold auto-inflammatory syndrome, Muckle-Wells syndrome, and neonatal onset multisystem inflammatory disease (Tschopp et al., 2003). Alterations in TLR-dependent responses are also associated with the progression of chronic inflammatory disorders such as allergic asthma, atherosclerosis, and arthritis (Basu and Fenton, 2004). Individual variations in TLR sequence promote differences in the regulation of signal strength and attenuation which can prolong responsiveness. Some evidence has shown that a deficiency in MyD88 function can be associated with the development of atherosclerosis, the generation of autoantibodies in SLE, and rheumatoid arthritis (Takeda and Akira, 2005).

Chemokines and chemokine receptors are a large family of at least 50 different chemokines and 19 different chemokine receptors that are also involved in a wide variety of disease states supporting the idea that recruitment of leukocytes is a critical component in regulation of the inflammatory response (Charo and Ransohoff, 2006). For example, dysfunctions in chemokine receptors are associated with different inflammatory disorders: CCR1 and CCR2 are associated with atherosclerosis, rheumatoid arthritis and multiple sclerosis; CCR3 with allergic asthma; CCR2 and CXCRX6 with atherosclerosis; and CXCR1 and CXCR2 are associated with inflammatory lung disease. Similarly, dysfunction in chemokines can also be correlated with different inflammatory disorders: CCL2 (a ligand for CCR2 also known as MCP-1α) are correlated with atherosclerotic plaques while CCL2, CCL3 (MIP-1α), CCL4 (MIP-1β), CCL5 (Rantes), and CCL8 are present in multiple sclerosis lesions.

Mutations in complement and complement receptor proteins also generate a range of disorders in the function of the innate immune system which are related to dysfunction in opsonization or lytic activity (Berliner et al., 2004). Deficiencies in C3, the major opsonin, results in recurrent pyogenic infections, particularly with encapsulated bacteria and patients may die from overwhelming sepsis. Deficiencies in early classical components, such as C1 and C4, do not usually predispose individuals to severe infections but are associated with increased incidence of autoimmune diseases such as SLE. Deficiencies in the later components of the cascade, which make up the MAC, do not usually promote high levels of morbidity although these patients do have a higher incidence of infections with *Neisseria*. Deficiencies in the complement receptors also contribute to immune disorders. Of the 8 different complement receptors currently described, congenital deficiencies in CR3 and CR4 (which are the result of a deletion in the shared β2 integrin subunit and are associated with Leukocyte adhesion deficiency in phagocytes) are the only ones described. Although congenital deficiency in the remaining complement receptors have not been described, acquired forms of deficiency, usually as a result of autoimmune disorders, are known (Berliner et al., 2004). For example, CR1 deficiency is associated with SLE, diabetic nephropathy, or preeclampsia and may account for increased infections in these patients.

It is interesting to note that dysfunction in a wide array of proteins and cell types that are part of the innate immune system are associated with various clinical disorders that result either in an increased susceptibility to pathogen infection or an increase in the incidence of chronic inflammatory conditions. This indicates that the appropriate balance in the regulation of an innate immune response is important for maintaining good health and those alterations in any of a large number of components can result in a shift in this balance resulting in clinically relevant consequences.

References

Acheson, D, Luccioli, S. Mucosal immune responses. *Best Practice and Research Clinical Gastroenterology,* 2004, 18:387-404.

Aderem, A, Underhill, DM. Mechanisms of phagocytosis in macrophages. *Ann Rev Immunol*, 1999, 17:593-623.

Aderem, A, Ulevitch, RJ. Toll-like receptors in the induction of the innate immune response. *Nature,* 2000, 406:782-787.

Ahmad-Nejad, P, Hacker, H, Rutz, M, Bauer, S, Vabulas, RM, Wagner, H. Bacterial CpG-DNA and lipopolysaccharides activate Toll-like receptors at distinct cellular compartments. *Eur J Immunol,* 2002, 32:1958-1968.

Akira, S., Uematsu, S, Takeuchi, O. Pathogen recognition and innate immunity. *Cell,* 2006, 124:783-801.

Alexander, C, Rietschel, ET. Invited review: Bacterial lipopolysaccharide and innate immunity. *J Endotoxin Res,* 2001, 7:167-202.

Balachandran, S, Roberts, P, Brown, L, Truong, H, Pattnaik, A, Archer, D, Barber, G. Essential role for the dsRNA-dependent protein kinase PKR in innate immunity to viral infection. *Immunity,* 2000, 13:129-141.

Banchereau, J, Steinman, RM. Dendritic cells and the control of immunity. *Nature,* 1998, 392:245-252.

Basu, S, Fenton, MJ. Toll-like receptors: function and roles in lung disease. *Am J Physiol Lung Cell Mol Physiol*, 2004, 286:L887-L892,

Berliner, N, Horwitz, M, Loughran, TP, Jr. Congenital and acquired neutropenia. *Hematology*, 2004, 1:63-79.

Biron, CA, Nguyen, KB, Pien GC, Cousens, LP, Salazar-Mather, TP. Natural killer cells in antiviral defense: function and regulation by innate cytokines. *Ann Rev Immunol*, 1999, 17:189-220.

Brinkmann, V, Reichard, U, Goosmann, C, Fauler, B, Uhlemann, Y, Weiss, DS, Weinrach, Y, Zychlinsky, A. Neutrophil extracellular traps kill bacteria. *Science,* 2004, 303:1532-1535.

Burg, ND, Pillinger, MH. The neutrophil: Function and regulation in innate and humoral immunity. *Clinical Immunology*, 2001, 99:7-17.

Caroll, MC. The role of complement and complement receptors in induction and regulation of immunity. *Ann Rev Immunol*, 1998,16:545-568.

Charo, IF, Ransohoff, RM. The many roles of chemokines and chemokine receptors in inflammation. *New Eng J Med*, 2006, 354:610-621.

Cooper, MA, Fehniger, TA, Caligiuri, MA. The biology of human natural killer-cell subsets. *Trends Immunol,* 2001, 22:633-640.

Dajani, R, Zhang, Y, Taft, PJ, Travis, SM, Starner, TD, Olsen, A, Zabner, J, Welsh, MJ, Engelhardt, JF. Lysozyme secretion by submucosal glands protects the airway from bacterial infection. *Am J Respir Cell Mol Biol*, 2005, 32:548–552.

Dale, D, Boxer, L, Liles, WC. The phagocytes: neutrophils and monocytes. *Blood*, 2008, 112:935-945.

Elias, PM. The skin barrier as an innate immune element. *Semin Immunopathol,* 2007, 29:3–14.

Fujita, T, Matsushita, M, Endo, Y. The lectin-complement pathway – its role in innate immunity and evolution. *Immunol Rev,* 2004, 198:185-202.

Ganz, T. Defensins: antimicrobial peptides of innate immunity. *Nature Rev Immunol,* 2003, 3:710-720.

Garlanda, C, Bottazzi, B, Bastone, A, Mantovani A. Pentraxins at the crossroads between innate immunity, inflammation, matrix deposition, and female fertility. *Ann Rev Immunol,* 2005, 23:337-366.

Gasque, P. Complement: a unique innate immune sensor for danger signals. *Mol Immunol,* 2004, 41:1089-1098.

Gorden, KKB, Qiu, X, Battiste, JJL, Wightman, PPD, Vasilakos, JP, Alkan, SS. Oligodeoxynucleotides differentially modulate activation of TLR7 and TLR8 by imidazoquinolines . *J Immunol*, 2006, 177:8164-8170.

Gordon, S. Pattern recognition receptors doubling up for the innate immune response. *Cell*, 2002, 111:927-930.

Greenberg, S, Grinstein, S. Phagocytosis and innate immunity. *Curr Opin Immun*ol, 2002, 14: 136-145.

Hajjar, AM, O'Mahony, DS, Ozinsky, A, Underhill, DM, Aderem, A, Klebanoff, SJ, Wilson, CB. Cutting Edge: Functional interactions between Toll-Like receptor (TLR) 2 and TLR1 or TLR6 in response to phenol-soluble modulin. *J Immunol,* 2001, 166:15-19.

Harding, CV, Song, R, Phagocytic processing of exogenous particulate antigens by macrophages for presentation by class I MHC molecules. *J Immunol,* 1994, 153:4925-4933.

Hart, SP, Smith, JP, Dransfield, I. Phagocytosis of opsonized apoptotic cells: roles for 'old-fashioned' receptors for antibody and complement. *Clin Exp Immunol,* 2004, 135:181–185.

Hayashi, F, Smith, KD, Ozinsky, A, Hawn, TR, Yi, EC, Goodlett, DR, Eng, JK, Akira, S, Underhill, DM, Aderem, A. The innate immune response to bacterial flagellin is mediated by Toll-like receptor 5. *Nature* 2001, 410:1099-1103.

Hawiger, J. Innate immunity and inflammation: a transcriptional paradigm. *Immunol Res* 2001, 23:99-109.

Heller, RU, Schena, M, Chai, A, Shalon, D, Bedilion, T, Gilmore, J, Woolley, DE, Davis, RW. Discovery and analysis of inflammatory disease-related genes using cDNA microarrays. *Proc. Natl. Acad. Sci. USA,* 1997, 94:2150–2155.

Inohara, N, Nunez G, NODS: intracellular proteins involved in inflammation and apoptosis. *Nature Reviews Immunology*, 2003, 3:371-382.

Jacobsen, E, Taranova, A, Lee, N, Lee, J. Eosinophils: Singularly destructive effector cells or purveyors of immunoregulation? *Journal of Allergy and Clinical Immunology*, 2007, 119:1313-1320.

Jiang, Z, Zamanian-Daryoush, M, Nie, H, Silva, AM, Williams, BRG, Li, X. Poly(dI-dC)-induced Toll-like Receptor 3 (TLR3)-mediated activation of NFkB and MAP kinase is through an interleukin-1 receptor-associated kinase (IRAK)-independent pathway employing the signaling components TLR3-TRAF6-TAK1-TAB2-PKR. *J Biol Chem,* 2003, 278:16713-16719.

Kimbrell, DA, Beutler, B. The evolution and genetics of innate immunity. *Nature Rev Genet,* 2001, 2:256-267.

Kobayashi, SD, Voyich, JM, Burlak, C, DeLeo, FR. Neutrophils in the innate immune response. *Arch Immunol Ther Exp*, 2005, 53:505–517.

Laskin, DL, Weinberger, B, Laskin, JD. Functional heterogeneity in liver and lung macrophages. *J Leuk Biol*, 2001, 70:163-170.

Leckstrom-Himes, JA, Gallin, JI. Immunodeficiency diseases caused by defects in phagocytes. *New Engl J Med*, 2000, 343:1703-1714.

Lin, WW, Karin, M. A cytokine-mediated link between innate immunity, inflammation, and cancer. J. *Clin Invest*, 2007, 117:1175-1187.

Medzhitov, R, Janeway, CA Jr. Decoding the patterns of self and nonself by the innate immune system. *Science*, 2002, 296:298 -300.

Medzhitov, R. Toll-like receptors and innate immunity. *Nature Rev Immunol*, 2001, 1:135-144.

Medzhitov, R, Biron, CA. Innate immunity. *Curr Opin Immunol*, 2003, 15:2-4.

Mekori, YA, Metcalfe, DD. Mast cells in innate immunity. *Immunol Rev*, 2002, 173:131-140, 2000.

Meylan, E, Tschopp, J, Karin, M. Intracellular pattern recognition receptors in the host response. *Nature*, 2006, 442:39-44.

Miller, SI, Ernst, RK, Bader, MW. LPS, TLR4 and infectious disease diversity. *Nature Rev Microbiol* 2005, 3:36-46.

Mizgerd, JP, Spieker, MR, Doerschuk CM. Early response cytokines and innate immunity: essential roles for TNF receptor 1 and type 1 IL-1 receptor during Escherichia coli pneumonia in mice. *J Immunol*, 2001, 166:4042-4048.

Nathan, C. Points of control in inflammation. *Nature*, 2002, 420:846-852.

Nauta AJ, Castellano G, Xu W, Woltman AM, Borrias MC, Daha MR, van Kooten C, Roos A. Opsonization with C1q and mannose-binding lectin targets apoptotic cells to dendritic cells.J Immunol. 2004 173:3044-3050.

Nishiya, T, DeFranco, AL. Ligand-regulated chimeric receptor approach reveals distinctive subcellular localization and signaling properties of the Toll-like receptor. *J Biol Chem*, 2004, 279:19008-19017.

Nürnberger, T, Brunner, F, Kemmerling, B, Piater, L. Innate immunity in plants and animals: striking similarities and obvious differences. *Immunological reviews*, 2004, 198:249-266.

Ozinsky, A, Underhill, DM, Fontenot, JD, Hajjar, AM, Smith, KD, Wilson, CB,
Schroeder,L, Aderem, A. The repertoire for pattern recognition of pathogens by the innate immune system is defined by cooperation between Toll-like receptors. *Proc Natl Acad Sci USA*, 2000, 97:13766-13771.

Palsson-McDermott, EM, O'Neill, LAJ. Building an immune system from nine domains. *Biochem Soc Trans*, 2007, 35:1437-1444.

Pepys, MB, Hirschfield, GM. C-reactive protein: a critical update. *J Clin Invest*, 2003,111:1805–1812.

Ravetch, JV, Kinet, JP. Fc Receptors. *Ann Rev Immunol*, 1991, 9:457-492.

Richer, MJ, Lavallee, DJ, Shanina, I, Horwitz, MS. Toll-like receptor 3 signaling on macrophages is required for survival following coxsackievirus B4 infection. *PLOS One*, 2009, 4:e4127.

Rus, H, Cudrici, C, Niculescu, F. The role of the complement system in innate immunity. *Immunol Res*, 2005, 33:103-112.

Sabroe, I, Read, RC, Wyte, MKB, Dockrell, DH, Vogel SN, Dower, SK. Toll-like receptors in health and disease: complex questions remain. *J Immunol*, 2003, 171:1630-1635.

Sahly, H, Keisari, Y, Crouch, E, Sharon, N, Ofek, I. Recognition of bacterial surface polysaccharides by lectins of the innate immune system and its contribution to defense against infection: the case of pulmonary pathogens. *Infection and Immunity*, 2008, 76:1322-1332.

Serbina, NV, Jia, T, Hohl, TM, Palmer, EG. Monocyte-mediated defense against microbial pathogens. *Ann Rev Immunol*, 2008, 26:421-452.

Stahl PD, Ezekowitz RAB The mannose receptor is a pattern recognition receptor involved in host defense. *Curr Opin Immunol*, 1998, 10:50-55.

Takahashi, M, Iwaki, D, Kanno, K, Ishida, Y, Xiong,J, Matsushita, M, Endo, Y, Miura, S, Ishii, N, Sugamura, K, Fujita, T, Mannose-binding lectin (MBL)-associated serine protease (MASP)-1 contributes to activation of the lectin complement pathway. *J Immunol*, 2008, 180:6132-6138.

Takeda, K, Akira, S. Toll-like receptors in innate immunity. *International Immunology*, 2005, 17:1-14.

Tlaskalova-Hogenova, H, Pankova R, Hudcovic, T, Tu, L, Cukrowska, B, Lodinova-adnikova, R, Kozakova, H, Rossmann, P, Bartova, J, Sokol, D, Funda, DP, Borovska, D, Rehankova, A, Sinkora, J, Hofman, J, Drastich, P, Kokevova, A. Commensal bacteria (normal microflora), mucosal immunity and chronic inflammatory and autoimmune diseases. *Immunol Lett*, 2004, 93:97-108.

Tschopp J, Martinon, F, Burns, K. NALPs: a novel protein family involved in inflammation, *Nature Reviews Molecular Cell Biology*, 2003, 4:95-104.

Triantafilou, M, Triantafilou, K. Lipopolysaccharide recognition: CD14, TLRs and the LPS-activation cluster. *Trends Immunol*, 2002, 23:301-304.

Wedermeyer, J, Tsai, M, Galli, SJ. Roles of mast cells and basophils in innate and acquired immunity. *Curr Opin Immunol*, 2000, 12:624-631.

Yamamoto, M, Sato, S, Hemmi, H, Sanjo, H, Uematsu, S, Kaisho, T, Hoshino, K, Takeuchi, O, Kobayashi, M, Fujita, T, Takeda, K, Akira, S. Essential role for TIRAP in activation of the signaling cascade shared by TLR2 and TLR4. *Nature*, 2002, 420:324-329.

Yang, S, Tamai, R, Akashi, S, Takeeuchi, O, Akira, S, Sugawara, S, Takada, H. Synergistic effect of muramyldipeptide with lipopolysaccharide or lipoteichoic acid to induce inflammatory cytokines in human monocytic cells in culture. *Infect Immunol*, 2001, 69:2045-2053.

Zanetti, M. Cathelicidins, multifunctional peptides of the innate immunity. *J Leuk Biol,* 2004, 75:39-48.

In: Molecular Aspects of Infectious Diseases
Editor: Mazen T. Saleh
ISBN: 978-1-61728-690-2

Chapter VII

Respiratory Infections

Marina Ulanova
Northern Ontario School of Medicine, West Campus, Lakehead University, MS 3006, 955 Oliver Road, Thunder Bay, ON P7B 5E1, Canada

Abstract

The respiratory system is consistently exposed to a variety of airborne pathogenic microorganisms. Colonization of the nasopharynx with pathogenic microorganisms is common, but the respiratory zone is sterile under the normal conditions; if the pathogenic microbes reach the respiratory zone pneumonia develops. The respiratory system has highly efficient defense mechanisms, i.e. the structural barriers, as well as innate and adaptive immunity. Pathogenic microorganisms have developed sophisticated molecular strategies to overcome the host defense mechanisms. In this chapter, molecular mechanisms involved in host-pathogen interactions during bacterial pneumonia, are illustrated on examples of two bacteria: *Streptococcus pneumoniae*, the most common cause of community-acquired pneumonia, and *Pseudomonas aeruginosa*, an important opportunistic pathogen causing pneumonia in the immunocompromised host, hospital-acquired pneumonia, and chronic pulmonary infections in cystic fibrosis patients. The major emphasis is on the complex interplay between bacterial virulence factors and host defense. The mechanisms of immune evasion by *S. pneumoniae* and *P. aeruginosa* are discussed in the context of the pathogenesis of pulmonary infections.

Introduction

The respiratory system anatomically and functionally consists of upper airways (nose and paranasal sinuses, pharynx, larynx), conducting airways (tracheal-bronchial tree), and the respiratory zone (respiratory bronchioles and alveoli), and it is the largest body area exposed to the outside environment. Indeed, each day approximately 9,700 L of air is inhaled and it is estimated that some 10,000 microorganisms, including pathogenic viruses, bacteria, and fungi, are inhaled daily by every city inhabitant. It is astonishing that with such a large load of pathogenic microorganisms, the respiratory infections do not permanently affect the human beings. Indeed, the respiratory zone, or the gas-exchange area, remains sterile, although the upper airways, especially the nasopharynx, are commonly colonized by commensal microorganisms.

The respiratory system is perfectly equipped to prevent infections at all anatomical levels, and each compartment has appropriate defense mechanisms. The destiny of inhaled microorganisms largely depends on their size: large microbes are usually trapped in the nose and the upper airways and then effectively removed with the mucociliary clearance; microbes < 5 μm can travel as far as to the alveoli, where they are phagocytosed by resident alveolar macrophages or by neutrophils recruited to the lung from bloodstream [1].

In the airway secretions, numerous antibacterial substances are present, such as lysozyme, lactoferrin, and peroxidase in the upper airways and collectins in the conducting airways and the respiratory zone (Table 7.1). Small cationic peptides (defensins) secreted by airway epithelia and neutrophils, have both direct bactericidal activities and abilities to induce complement activation, chemokine production, and T-cell proliferation [2]. Collectins, i.e. mannose-binding lectin (MBL) and surfactant proteins A and D contain C-type lectin domains and bind various microorganisms acting as opsonins. The resulting enhanced phagocytosis leads to bacterial clearance. MBL also activates complement that results in bacterial lysis [3, 4]. Pentraxins, i.e. C-reactive protein, serum amyloid P, and the long pentraxin PTX3 contribute to lung innate defense via opsonization and complement activation [5]. The complement system is important in host defense via recruitment and activation of phagocytic cells, opsonization of bacteria, and direct lysis of microorganisms [6].

Table 7.1. Immune defense of the respiratory system.

Complement
• Cellular factors:
Phagocytic cells: alveolar macrophages, neutrophils, dendritic cells
NK cells
Mast cells
Epithelial cells
Innate immune receptors
Adaptive immunity
• Circulating IgG, IgM, IgE antibodies
• Secretory IgA antibodies
• CD4+ T cells

Abbreviations: Ig, immunoglobulin; MBL, mannose-binding lectin; NK, natural killer

Cell types involved in innate immune responses in the respiratory system include phagocytes, such as alveolar macrophages, neutrophils, and dendritic cells capable of ingesting pathogens. Phagocytosis is followed by destruction of microbes with lysosomal enzymes and reactive oxygen metabolites. Phagocytosis by macrophages and dendritic cells is critical for antigen presentation and initiation of adaptive immune responses. Natural killer (NK) cells are capable of both direct killing of infected host cells and large pathogens (e.g. parasites), and of immunoregulation. During the infectious process, mast cells and epithelial cells release a plethora of biologically active molecules, mediators of inflammatory responses, regulating recruitment of blood-derived leukocytes and activating other cells in the lung. Epithelial cells are especially important in orchestrating of innate immune and inflammatory responses in the lung because they are the most prevalent cell type in the respiratory system, lining the whole surface area of the airways [7, 8].

Adaptive immune responses are mediated by specific T lymphocytes and antibodies against antigens of pathogenic microorganisms. Circulating immunoglobulin G (IgG) and IgM antibodies are capable of opsonising bacteria consequently enhancing their phagocytosis and lysis by complement. These antibodies also neutralize virus particles and bacterial toxins and mediate antibody-dependent cellular cytotoxicity, which helps to destroy infected cells expressing microbial antigens as well as large parasites. IgE antibodies, although involved in allergic responses, have a protective effect against parasites as they cause degranulation of mast cells with the release of large amounts of toxic substances to kill parasites. Secretory IgA produced locally in mucosal tissues, bind pathogens and prevent their adhesion to susceptible cells, as well as neutralise viruses and toxins. Cell-mediated immune responses are critical in the regulation of antibody production and activation of macrophage function; the cytotoxic T lymphocytes destroy infected host cells displaying antigenic epitopes of intracellular pathogens on their surface.

Activation of cells of the innate immune system requires recognition of conserved microbial molecules by the pattern-recognition receptors (PRR), such as toll-like receptors (TLR) and nucleotide-binding oligomerization domain (NOD)-like receptors [9] (Table 7.2). The best studied are TLR; at least 10 mammalian TLR exist; these receptors recognize a wide range of pathogen-associated molecules. The microbial recognition leads to activation of TLR-mediated cellular signalling pathways, which regulate both innate and adaptive immunity. The common TLR adaptor protein, myeloid-differentiation primary-response protein 88 (MyD88) is critically important in host defense. Genetic defects in either MyD88 or a signalling molecule acting downstream of MyD88, interleukin-1-receptor-associated kinase (IRAK-4) lead to a severe immunodeficiency, which manifests as an enhanced susceptibility to invasive bacterial infections, especially ones caused by pneumococci [10, 11].

Pathogenic microorganisms have developed sophisticated molecular strategies to overcome the host defense mechanisms. Some microbes use host cellular receptors for adhesion to susceptible cells and their invasion. The best-studied example of bacteria, which are able to directly bind integrin receptors of the host cells and exploit integrin-mediated signalling mechanisms, resulting in bacterial invasion, is the enteric pathogen *Yersinia pseudotuberculosis* [12]. *Neisseria gonorrhoeae*, which infects mucosal tissues of the genitourinary tract, uses the type IV pili and colony opacity-associated (Opa) proteins to adhere to and invade epithelial cells via specific engagement of carcinoembryonic-antigen-related cell adhesion molecules [13]. *Bordetella pertussis*, the causative agent of whooping cough, uses the filamentous hemagglutinin to bind host cell integrin receptors [14]. Many microbes are capable of evading the host's defense either by escaping immune recognition or destroying critical protective molecules, e.g. *Staphylococcus aureus* or *Mycobacterium tuberculosis* [15, 16]. Moreover, some microorganisms are capable of exploiting intracellular signalling pathways of infected cells to their benefit, e.g. *Pseudomonas aeruginosa* [17, 18]. The outcomes of pathogen-host interactions largely depend on the balance between bacterial virulence factors and host defense mechanisms and can range from rapid clearance of the infectious agent or its asymptomatic carriage in the upper airways, to severe pneumonia or life-threatening systemic septicaemia.

Table 7.2. Major families of innate immune receptors and their ligands.

Receptor Family	Ligands or Activators
Toll-like receptors: TLR1-12	Bacterial LPS, lipoprotein, lipoteichoic acids, lipoarabinomannan, peptidoglycan, flagellin, *Pseudomonas* exotoxin S, bacterial and viral DNA, viral RNA, fungal zymosan and mannan, parasitic phospholipids, neutrophil elastase
Nod-like receptors: NOD1-2, RIG-I, MDA5, DAI, NALP subfamily	Bacterial peptidoglycans, viral nucleic acids, extracellular ATP, uric-acid crystals, bacterial flagellin
C-type lectins: dectin 1, MR, DC-SIGN	Fungal β-glucan, mannose oligosaccharides, HIV gp120, lysosomal hydrolases, fungi, viruses
Scavenger receptors: SRA, CD36, MARCO	Apoptotic cells, bacterial diacyl lipids
Integrins: α5β1, CR3, CR4	FnBP, invasin, FHA, complement activation products, LPS
Complement and complement receptors: C3, gC1qR, C5aR	Immune complexes, apoptotic cells, bacterial capsular polysaccharides, C1q, C5a

Abbreviations: C, complement, CD, cluster of differentiation, CR, complement receptor, DAI, DNA-dependent activator of interferon-regulatory factors, DC-SIGN, dendritic cell-specific intercellular adhesion molecule-3-grabbing non- integrin, FHA, filamentous hemagglutinin, FnBP, fibronectin binding protein, HIV, human immunodeficiency virus, LPS, lipopolysaccharide, MARCO, macrophage receptor with collagenous structure, MDA, melanoma differentiation-associated gene, MR, mannose receptor, NALP, NACHT-, LRR- and pyrin-domain-containing proteins, Nod, nucleotide-binding oligomerization domain, RIG, retinoic-acid-inducible gene, SR, scavenger receptor

Pulmonary Infections

Despite the fact that microorganisms smaller than 5 μm can travel with inhaled air straight to the alveoli, the gas-exchange area of the respiratory system is sterile. Any infection or inflammation of this area is highly detrimental as it would affect the crucial physiological function of the respiratory system, i.e. to provide oxygen in exchange for carbon dioxide.

Pneumonia can be caused by a variety of pathogenic organisms, including bacteria, viruses, fungi and protozoa. Microorganisms most commonly causing community-acquired pneumonia include *Streptococcus pneumoniae*, *Haemophilus influenzae,* and *Staphylococcus aureus. Moraxella catarrhalis, Legionella pneumophila,* and *Klebsiella pneumoniae* are also important etiological agents of community-acquired pneumonia, although they are less common. Microorganisms such as *Mycoplasma pneumoniae, Chlamydia* spp. (*C. pneumoniae, C. psittaci, and C. trachomatis*), *Coxiella burnetti* (cause of Q fever), respiratory syncytial, parainfluenza, influenza viruses, adenovirus, and SARS coronavirus cause atypical community-acquired pneumonia. The spectrum of microorganisms causing pneumonia in hospital settings (hospital-acquired, or nosocomial pneumonia) is quite different and is dominated by Gram-negative rods belonging to Enterobacteriaceae (*Klebsiella* spp., *Serratia marcescens, Escherichia coli*) and *Pseudomonas* spp.; among Gram-positive bacteria, *Staphylococcus aureus* is most common. Pneumonia caused by breathing in foreign material (aspiration pneumonia) is mostly associated with anaerobic oral flora (*Bacteroides, Prevotella, Fusobacterium, Peptostreptococcus*), mixed with aerobic bacteria *(Streptococcus pneumoniae, Staphylococcus aureus, Haemophilus influenzae, Pseudomonas aeruginosa).* In immunocompromised people, e.g. AIDS patients or individuals with primary immunodeficiencies, pneumonia is often caused by unusual opportunistic pathogens, such as *Cytomegalovirus, Pneumocystis jiroveci* (formerly *carinii*), *Mycobacterium avium, Aspergilla* spp., *Candida* spp.; although pneumonias caused by "usual" bacterial or viral organisms are also present [19].

Pneumonia occurs when the balance between the virulence factors of the pathogen and host defense is shifted in favour of the pathogen. This may happen when the normal defense mechanisms in the respiratory system are insufficient to prevent the infectious process caused by a highly virulent organism. Indeed, some pathogens are capable to significantly compromise the host defense, e.g. SARS coronavirus [20]. Pneumonia may develop as a result of defects in the normal clearing mechanisms of the respiratory system caused, for example, by an injury to the mucociliary clearance or suppression of the cough reflex. Any damage to the functional capabilities of the immune system, such as primary and secondary immunologic deficiencies, chronic diseases, treatment with immunosuppressive agents or leucopenia can predispose to the development of pneumonia. Insufficient immune defense against some pathogens may be a result of a lack of specific antibodies in otherwise normal individuals due to an immature immune system (e.g. absence of antibodies against bacterial capsular polysaccharides in children below 2 years of age) or due to the lack of previous exposure to protective antigens of the pathogen via asymptomatic carriage or immunization [21].

In the next sections, molecular mechanisms involved in host-pathogen interactions during bacterial pneumonia, will be illustrated on examples of two bacterial pathogens:

Streptococcus pneumoniae, the most common cause of community-acquired pneumonia, and *Pseudomonas aeruginosa*, an important opportunistic pathogen causing pneumonia in the immunocompromised host, hospital-acquired acute pneumonia as well as chronic pulmonary infections in cystic fibrosis patients.

Streptococcus Pneumoniae

The leading cause of community-acquired pneumonia, *Streptococcus pneumoniae* (pneumococcus) is a Gram-positive encapsulated microorganism, frequently colonizing the upper respiratory tract, most commonly, in young children (carriage rate in Western Europe and U.S. is 20-50%) [22]. If *S. pneumoniae* overcomes the host defense mechanisms it is able to cause otitis media, pneumonia, meningitis and septicemia. Pneumococcal capsular polysaccharides are the major virulence factor; in absence of specific antibodies, the capsule provides efficient protection against opsonophagocytosis [23].

There are at least 90 capsular types of *S. pneumoniae* different in their serological properties as well as in virulence. In the U.S. and Western Europe, 23 serotypes cause 85-90% of pneumococcal infections and these are included into the current polysaccharide vaccine. Since children below 2 years of age do not mount protective antibody responses to pure bacterial capsular polysaccharides, for pediatric immunization, protein-polysaccharide conjugate vaccine containing capsular polysaccharide antigens of 7 most common *S. pneumoniae* serotypes is used [24].

In the process of development of pneumococcal pneumonia, the following stages of microbial pathogenesis can be identified: upper airway colonization – invasion – dissemination – tissue damage – immune response/inflammation – resolution. These stages are characterized by a different balance of bacterial virulence factors/host defense mechanisms as well as by specific molecular strategies used by the pathogen. Below, the mechanisms of microbial pathogenesis are discussed in the context of interactions between pneumococcal virulence factors and the host (Table 7.3).

Colonization of Upper Airways

Asymptomatic carriage of *S. pneumoniae* in the nasopharynx is frequently found in normal individuals, especially in young children. In the upper respiratory tract, *S. pneumoniae* behave as commensal bacteria, being under control of host immune defense mechanisms. However, clinical infection can occur when pneumococci spread beyond this niche into normally sterile sites of the respiratory system, i.e. alveoli. Any damage to the epithelial barriers of the upper airways, such as one caused by respiratory viruses, can predispose to local pneumococcal infection, e.g. otitis media. Under certain circumstances, especially in young children, pneumococci are able of penetrating the nasopharyngeal mucosae and gaining access to the systemic circulation via the lymphatic system. This will lead to bacteremia with potential development of focal infections, such as meningitis, arthritis, or endocarditis. Hence, colonization of the upper respiratory tract represents the initial critical step in the pathogenesis of pneumonia as well as other clinical forms of pneumococcal infection, i.e. otitis media, meningitis, and septicemia [25].

Colonization of the mucosal surfaces of the upper airways is initiated upon adhesion of pneumococci to the respiratory epithelium. Pneumococci possess numerous adhesive proteins and utilize several host receptors for adhesion to epithelial cells; several pneumococcal adhesins serve both for adherence to and invasion of host cells.

Phosphorylcholine, a specific component of pneumococcal cell wall mediates adherence to the G-protein-coupled receptor for platelet-activating factor (PAFr) and activates host cell signaling through this receptor [26]. Recognition of phosphorylcholine by PAFr is especially important during the interaction of pneumococci with activated host cells and is involved in bacterial transcytosis through the blood-brain barrier [27]. Some important interactions of pneumococci with host cells are regulated by the *pce* gene, which encodes for a phosphorylcholine esterase (Pce), an enzyme removing phosphorylcholine residues from the cell wall teichoic and lipoteichoic acids [28, 29].

Table 7.3. Balance between virulence factors and host defense in the pathogenesis of pneumococcal pneumonia.

Mechanisms of pathogenesis	Role of virulence factors	Mechanisms of host defense	Microbial evasion	Predisposing host factors
Colonization of upper airways	Pneumococcal adhesins Use of host cell receptors and ECM	Mucociliary clearance Secretory IgA Bactericidal factors of airway secretions	Polysaccharide capsule IgA1 protease	Defect of cilia and mucus function Viral infections Smocking CF
Invasion and dissemination in the lung	Exploitation of the polymeric Ig transport system Activation of the plasminogen-plasmin system Effect of toxins (pneumolysin, neuraminidases, hyaluronidase, autolysin)	Opsonophago-cytosis IgG and IgM antibodies Complement activation	Polysaccharide capsule	Lack of capsule-specific antibodies (infants, elderly) PID:hypo-gammaglobulinemia, complement deficiency Impaired PMN migration (alcohol, anaesthetics, corticosteroids)
Immune response and inflammation resulting in resolution of the infectious process	PAMP: peptidoglycan, pneumolysin, lipoteichoic acid, lipoproteins, bacterial DNA	Recognition by PRR and activation of innate immunity Cytokine release Antigen presentation and adaptive immune response Complement activation CRP PMN recruitment	Inhibition of complement system Inhibition of recognition by DC	I RAK-4 or MyD88 deficiency PID:hypo-gammaglobulinemia, complement deficiency Asplenia

Abbreviations: CF, cystic fibrosis, CRP, C-reactive protein, DC, dendritic cells, ECM, extracellular matrix, Ig, immunoglobulin, IRAK-4, interleukin-1-receptor-associated kinase, MyD88, myeloid-differentiation primary-response protein 88, PAMP, pathogen-associated molecular pattern, PID, primary immunodeficiency, PMN, polymorphonuclear phagocyte, PRR, pattern-recognition receptor

Pneumococcal surface adhesin A (PsaA) is a 37-kDa lipoprotein and a member of an ATP-binding cassette (ABC) transporter complex for manganese. This molecule interacts with E-cadherin [30], the transmembrane glycoprotein, which forms calcium-dependent associations between cells [31]. Such interaction is mediated by a 28 amino acid long functional epitope of PsaA. Binding of pneumococci to nasopharyngeal epithelial cells via this epitope activates the host cells facilitating bacterial adhesion [32].

Some strains of *S. pneumoniae* possess adhesive pili-like appendages encoded by the *rlrA* pathogenicity islet. These structures enhance both initial bacterial adhesion, which is important for colonization, and the ability of pneumococci to cause invasive disease [33].

A surface-exposed lipoprotein of *S. pneumoniae*, cyclophilin SlrA contributes to adherence and invasion of host cells, especially during colonization of the upper airways; however, specific molecular mechanisms involved in these effects remain unknown [34].

S. *pneumoniae* is also capable of binding several extracellular matrix glycoproteins, such as trombospondin-1 (TSP1), fibronectin (Fn), vitronectin, and keratin 10. Interaction of pneumococci with TSP1 is mediated by bacterial peptidoglycan and promotes colonization of host cells. As TSP1 binds cellular receptors, e.g. β1 and β3 integrins, this extracellular glycoprotein may act as a molecular bridge linking bacteria with host cells [35].

Pneumococci bind vitronectin via specific interaction with the heparin-binding sites of this adhesive glycoprotein. Vitronectin is a ligand for αvβ3 integrin, and pneumococci use this receptor for adherence and internalization. Vitronectin-mediated invasion into host cells requires integrin-linked kinase, a signaling molecule essential for integrin-dependent cytoskeleton rearrangement, as well as phosphatidylinositol 3-kinase (PI3K) and protein kinase B (PKB, or Akt). Via this pathway, pneumococci exploit the vitronectin-αvβ3 integrin complex as a mechanism for invasion [36]. Similar mechanisms of bacterial exploitation of integrin-extracellular matrix molecule interaction involving other adhesive glycoproteins, such as Fn, laminin, collagens, etc are common among other respiratory pathogens, e.g. *Staphylococcus aureus*, *Streptococcus pyogenes*, and *Bordetella pertussis* [37].

A key virulence determinant of S. *pneumoniae* is the pneumococcal adherence and virulence factor A (PavA), which is an Fn binding protein structurally homologous to the Fn binding proteins of other pathogenic bacteria. This molecule is essential for the ability of pneumococci to cause invasive disease, such as meningitis [38, 39].

A pathogenicity island-encoded pneumococcal serine-rich repeat protein (PsrP) involved in bacterial adhesion to host cells, binds to keratin 10 on the surface of lung epithelial cells [40].

Invasion and Dissemination

As a result of adhesion, pneumococci can invade airway epithelial cells and this process can promote dissemination of microorganisms. Pneumococci can be also aspirated into the lower airways and consequently reach the smaller bronchi, bronchioles and alveoli. Pneumococcal invasion into alveolar epithelial cells will lead to further dissemination of bacteria, including penetration of the alveolar-endothelial membrane in the gas-exchange area causing bacteraemia. Pneumococcal invasion is a complex molecular process and is mediated by several virulence factors. Here again the bacteria demonstrate an amazing adaptability to the host and abilities to successfully exploit the host defense mechanisms.

As mucosal surfaces normally create significant barriers for bacterial penetration, pneumococci use sophisticated molecular mechanisms to cross epithelial cells and disseminate into the host tissues. Following initial adhesion to the surface of the airway epithelial cells, bacteria migrate through the epithelium either via an intracellular route, from the apical to basolateral surface, or transit between the adjacent cells, destroying tight junctions and making pores in the basement membrane.

The pneumococcal surface protein C (PspC), also known as choline-binding protein A (CbpA) or SIgA-binding protein (SpsA), mediates both bacterial adherence and invasion via direct interaction with the ectodomain of the polymeric immunoglobulin receptor (pIgR), known as the secretory component (SC) [41]. Specific binding between PspC and SC involves hexapeptide motifs located in the N-terminal repeat domains R1 and R2 of PspC and Ig-like ectodomains D3 and D4 of the SC [42, 43]. This molecular mechanism uses normal eukaryotic cellular machinery of transporting polymeric immunoglobulins, i.e. IgA dimers and IgM pentamers, across the epithelial barriers that is essential for mucosal immunity. Indeed, secretory IgA antibodies and to a certain degree, secretory IgM, are capable to prevent adhesion of bacteria to epithelial cells as well as directly kill certain pathogens. However, pneumococci are able to exploit this cellular transport mechanism, and as a result, successfully translocate across a mucosal barrier. Pneumococcal invasion via the PspC-pIgR interaction is mediated by the rearrangement of host cell cytoskeleton and involves the activation of complex intracellular signalling machinery of infected cells. In particular, this process requires the activity of a small GTPase Cdc42, PI3K, and PKB [44].

For invasion and dissemination, S. *pneumoniae* successfully exploit the host plasminogen-plasmin proteolytic system as the bacteria bind plasminogen using several surface receptors. Pneumococcal plasminogen-plasmin receptors include the glycolytic enzymes enolase and glyceraldehyde-3-phosphate dehydrogenase (GAPDH), as well as choline-binding protein E (CBPE) [45-47]. The crystal structure of alpha-enolase from *S. pneumoniae* has been resolved and the nine residue plasminogen-binding motif on the surface of octameric alpha-enolase identified as the primary site of interaction between alpha-enolase and plasminogen [48]. Also, the phosphorylcholine esterase domain of CBPE interacts with the plasminogen kringle domains involving several lysine residues [49]. Through these receptors, pneumococci recruit the proenzyme plasminogen. Its subsequent activation into plasmin, a broad-spectrum serine protease results in tissue barrier degradation due to hydrolysis of the extracellular matrix components. Proteolysis of the cell junction components, such as cadherin, especially affects the vascular endothelium. This mechanism is essential in bacterial transmigration across the basement membrane and promotes pneumococcal dissemination in the host tissues [47, 49].

Pneumococci express a potent cholesterol binding cytotoxin, pneumolysin, which is able to make pores in cell membranes causing rapid cell lysis or apoptosis. Pneumolysin is released as a soluble monomer, binds to cholesterol-containing membranes of target cells, assembles into large oligomeric rings, and forms pores causing cell death [50]. Via this mechanism, pneumolysin can facilitate bacterial migration through tissue barriers. In addition, sublytic concentrations of pneumolysin cause rapid activation of Rho and Rac GTPases followed by formation of actin stress fibers, filopodia, and lamellipodia providing another example of bacterial manipulation of the host signaling mechanisms [51]. Indeed, small GTPase modulation by bacteria contributes to the pathogenesis alleviating microbial penetration through cellular barriers and compromising host defense mechanisms [52]. In an experimental model, pneumolysin was demonstrated to be a significant factor contributing to acute lung injury in the course of severe pneumococcal pneumonia because this toxin can impair pulmonary microvascular barrier function and cause severe pulmonary hypertension [53].

Immune Defense and Inflammation

Some important virulence factors, e.g. peptidoglycan of *S. pneumoniae* are recognized by the host innate immune receptors TLR, NOD-like receptors, CD14, and LPS binding protein and consequently trigger immune defense mechanisms [54]. Pneumococcal peptidoglycan is recognized by both cell surface (TLR2) and intracellular receptors (NOD2); the latter receptor recognizes muramyl dipeptide, a common fragment of bacterial peptidoglycans [55]. TLR2 also recognizes the lipoteichoic acid and lipoproteins, and TLR2-deficient mice show delayed pneumococcal clearance [56]. Pneumolysin is recognized by TLR4 and subsequently triggers macrophage inflammatory responses critical for resistance to pneumococcal infection [57]. Intracellular TLR9 detects bacterial DNA and this is important during early stages of pneumococcal infection [58].

Activation of innate immune receptors is essential for the initiation of adaptive immune responses, because the signals generated as a result of TLR activation by microbial products, provide co-stimulation of cells of the adaptive immune system. Adaptive immunity to *S. pneumoniae* is characterized by the production of specific antibodies directed against capsular polysaccharide antigens. Polysaccharide antigens induce T-cell independent immune response as they cause immunoglobulin receptor cross-linking on the surface of B cells, rather than recognition by the T cell receptor in the context of the MHC molecules [59]. The following activation, differentiation and proliferation of B cells specific to the polysaccharide antigens lead to the production of protective antibodies of IgM, IgG2, and, to a lesser extent, of IgG1 isotype. The antibodies mediate opsonophagocytosis of pneumococci and as a result, neutrophils and macrophages engulf bacteria and kill them intracellularly via combined effects of lysosomal enzymes and reactive oxygen metabolites. Antibody response to capsular polysaccharides partially involves T-cell help, i.e. via the interaction between CD40 ligand on the surface of activated CD4+ T cells and CD40 expressed by B cells [60]. NK T cells play a key role in the innate phase of host defense against pneumococcal infection [61].

Adaptive immune response to protein antigens of *S. pneumoniae*, such as pneumolysin, PspA, and PspC also develops during the infectious process [62-64]. During pneumococcal lung infection, an early accumulation of CD4+ T cells is dependent on pneumolysin and critically important for protective host response [63]. Antigen presentation of pneumococcal antigenic peptides to T cells in the context of MHC class II molecules is carried out by dendritic cells and macrophages. It was demonstrated that alveolar macrophages phagocytose pneumococci, migrate along the lymphatic vessels and transport the antigenic material to lung draining lymph nodes where priming of naïve T cells takes place [65].

The activation of complement system by immune complexes composed of pneumococcal antigens and specific antibodies, leads to bacterial lysis. In addition, some innate immune factors, such as C-reactive protein (CRP), various lectins, and natural anti-carbohydrate IgM antibodies act as opsonins and upon deposition on the bacterial surface facilitate phagocytosis by neutrophils and macrophages [66]. The major component of the acute phase response, CRP specifically binds to phosphorylcholine and, following this binding, activates the classical complement pathway [67]. CRP is also able to inhibit pneumococcal adherence mediated by the interaction of phosphorylcholine with PAFr [68].

Development of innate immune response is associated with inflammation. Activation of resident alveolar macrophages and other cells in the lung, i.e. epithelial and mast cells, by microbial products results in production of pro-inflammatory cytokines and chemokines, which participate in innate immune responses. Among these secreted products are interleukin (IL)-1, IL-6, tumor necrosis factor (TNF)-α, IL-12, IL-18, interferon (IFN)-γ, etc. [66]. TNF-α is particularly important; its inhibition dramatically impairs the host defense against pneumococci in experimental models [69]. Inflammatory responses triggered by these molecules are characterized by a massive influx of leukocytes, including polymorphonuclear phagocytes. Neutrophil recruitment is regulated by many inflammatory mediators, including chemoattractants produced as a result of complement activation, chemokines, and galectin 3, which is a soluble host adhesion molecule [70-72].

Despite its critical role in infection clearance, the host inflammatory response can also compromise the barrier function of airway epithelial cells and facilitate transepithelial migration of pneumococci. Innate immune recognition of pneumococcal lipoteichoic acid by epithelial TLR2 results in the activation of intracellular signaling, i.e. p38 mitogen-activated protein kinase (MAPK) and transforming growth factor (TGF)-β mediated pathways, leading to the disruption of the epithelial barriers [73].

Immune Evasion

Several pneumococcal virulence factors mediate immune evasion. PavA, in addition to acting as the bacterial adhesin, protects pneumococci against recognition and phagocytosis by dendritic cells [74]. Another example is the polysaccharide capsule, which is the major factor preventing opsonophagocytosis of pneumococci in a non-immune organism. The capsule is also important in the process of initial pneumococcal colonization as it prevents entrapment of bacteria in the mucus of the upper airways. Since the majority of pneumococcal capsular polysaccharides are negatively charged, the capsule inhibits binding of pneumococci to mucus, which is also negatively charged, via electrostatic repulsion. As a result, pneumococci escape mucociliary clearance and are able to reach host cell receptors on the epithelial surface [75]. Because highly hydrophilic and anionic properties of the capsule also interfere with pneumococcal adherence to the cell surface, bacteria undergo phase variation and decrease capsule size during the initial stages of colonization. During later stages of the infectious process when pneumococci invade the host, they increase the capsule size again and thus escape immune defense mechanisms [76, 77]. Hence, pneumococci demonstrate great flexibility and adaptabilities in their interactions with the host.

Several pneumococcal products interfere with the complement system. The choline-binding pneumococcal surface protein A (PspA) protects pneumococci against the host immune defense via several mechanisms. PspA inhibits both classical and alternative complement pathways by interfering with C3 activation and C3b deposition [78]. In addition, PspA protects the bacteria against the bactericidal effects of apolactoferrin, which is an iron-depleted form of lactoferrin, possibly via blocking its active site [79].

In addition to its interaction with SC, PspC binds the complement factor H, a critical regulatory protein of the alternative pathway, although this time via a different part of the N-terminal alpha-helical region [80]. Because factor H is constitutively expressed by airway epithelial as well as endothelial cells, this mechanism promotes adhesion of pneumococci to the host cells [81]. Moreover, binding of factor H by PspC also mediates immune evasion of the pathogen. Indeed, factor H attached to the bacterial surface inhibits complement activation and therefore interferes with complement-mediated killing of pneumococci [82]. Expression of PspC by pneumococci of serotype 3 explains their particular resistance to phagocytosis [83, 84]. The PspC-like protein Hic (factor H-binding inhibitor of complement) of *S. pneumoniae* is also capable of binding factor H [83].

Pneumococci secrete zinc metalloprotease, which specifically targets human immunoglobulin A, i.e. IgA1 protease. This enzyme is an important virulence factor of *S. pneumoniae*; it is highly specific for prolyl-threonyl and prolyl-seryl bonds in the hinge region of human IgA1, the major isotype of immunoglobulins on mucosal surfaces [85]. IgA1 proteases of *S. pneumoniae* as well as of some other pathogenic bacteria destroy IgA1-antibody and consequently inhibit both its direct bactericidal effect and its ability to block microbial adhesion to epithelial cells [86]. In addition, pneumococcal IgA1 protease modifies IgA1 antibody so that it promotes bacterial adherence to epithelial cells. Specifically, Fab fragments of IgA antibody generated by removal of Fc fragments by IgA1 protease attach to bacteria and consequently alter the physical properties of bacterial cell surface. As a result, anti-adhesive effects of negatively charged capsular polysaccharide are blocked and pneumococci attach to the surface of airway epithelial cells [87]. This is a good example of exploitation of host defense mechanisms for the pathogen benefits.

Resolution of pulmonary inflammation caused by *S. pneumoniae* involves phagocytic clearance of apoptotic neutrophils, which accumulate in the lung parenchyma during acute pneumonia. This process is accomplished by alveolar macrophages; these cells also undergo apoptosis [88]. These functions of macrophages are controlled by the direct cytotoxic effect of $\gamma\delta$T cells, which help restore macrophage numbers in the lung to homeostatic levels and hence prevent inappropriate inflammation [89]. The roles of major virulence factors of *S. pneumoniae* in the pathogenesis are summarized in Table 7.4.

Table 7.4. Major virulence factors of Streptococcus pneumoniae and their role in the pathogenesis of pneumococcal infection.

Virulence factors	Molecular mechanisms of action	Role in the pathogenesis
Polysaccharide capsule	Prevents entrapment in the mucus and inhibits phagocytosis	Colonization of mucosal surfaces Immune evasion
Peptidoglycan	TLR2 and NOD2 ligand Binds TSP1	Innate immune response Inflammation Adhesion to host cells
Phosphorylcholine	Binds PAFr in epithelial cells and activates signalling via this receptor	Colonization of the nasopharynx Transcytosis through the blood-brain barrier
PsaA	Binds E-cadherin	Colonization of the nasopharynx
PavA	Binds fibronectin Prevents phagocytosis by DC	Invasion Immune evasion
Pili	Mediate adherence	Colonization and invasion

Table 7.4 (Continued)

Lipoteichoic acid	TLR2 ligand	Inflammation
PsrP	Binds keratin 10	Colonization
PspC (CbpA or SpsA)	Binds human pIgR and exploits the pIg transport mechanism Binds factor H and C3	Colonization and invasion Translocation across mucosal barriers Immune evasion
PspA	Inhibits complement activation Blocks the active site of apolactoferrin	Immune evasion
Pneumolysin	Cytolytic effect TLR4 ligand	Disruption of tissue barriers Dissemination Inflammation
Plasminogen-plasmin receptors: enolase, GAPDH, CBPE	Plasminogen activation	Disruption of tissue barriers Transmigration across the basement membrane Dissemination in tissues
IgA1 protease	Degrades IgA1 antibody and binds Fab fragments of IgA	Immune evasion

Abbreviations: C3, complement factor 3, CbpA, choline-binding protein A, CBPE, choline-binding protein E, DC, dendritic cella, GAPDH, glyceraldehyde-3-phosphate dehydrogenase, NOD, nucleotide-binding oligomerization domain, PAFr, platelet-activating factor receptor, PavA, pneumococcal adherence and virulence factor A, pIgR, polymeric immunoglobulin receptor, PsaA, pneumococcal surface adhesin A, PspA, pneumococcal surface protein A, PspC, pneumococcal surface protein C, PsrP, pneumococcal serine-rich repeat protein, SpsA, SIgA-binding protein, TLR, toll-like receptor, TSP1, trombospondin-1

PSEUDOMONAS AERUGINOSA

Clinical Significance

Pseudomonas aeruginosa is a ubiquitous microorganism commonly present in various environments, such as soil, water, plants, and animals. *P. aeruginosa* is a Gram-negative motile aerobic rod, which has a large genome and possesses remarkable abilities of adaptation to different environments. Due to its minimal nutritional requirements *P. aeruginosa* often grows in hospital settings including plumbing systems and inhalation equipment.

P. aeruginosa is an opportunistic pathogen and does not cause disease in humans with normal immune defense mechanisms. In the normal, immunocompetent host, natural defense mechanisms of the respiratory system are sufficient to prevent the development of infectious process. However, this microorganism causes severe life-threatening infections in individuals with impaired host defense. Patients with primary immunodeficiencies, such as primary ciliary dyskinesia [90] and leukocyte adhesion deficiency [91], secondary immunodeficiencies in hematopoietic stem cell transplant recipients [92], as well as ones caused by HIV infection [93], burns, hematopoietic malignancies, immunosuppressive therapy (e.g. neutropenia in cancer patients following chemotherapy) [94, 95], etc are both highly sensitive to *P. aeruginosa* pneumonia and prone to developing systemic septicemia caused by this microorganism. *P. aeruginosa* is the major cause of ventilator-associated pneumonia (VAP) in intensive care unit (ICU) patients with high mortality rate [96, 97]. Among other pulmonary conditions, *P. aeruginosa* is common in chronic obstructive pulmonary disease (COPD), especially in mechanically ventilated patients with acute exacerbations of COPD [98]. *P. aeruginosa* expresses a wide range of virulence factors, such as the type III secretion system, endotoxin, exotoxins, and various enzymes, which are able to significantly compromise host defense and induce potent activation of inflammatory responses [99] (Table 5). This pathogen is extremely capable of developing antibiotic resistance, making treatment difficult and often unsuccessful [100].

Table 7.5. Virulence factors of Pseudomonas aeruginosa.

Cell-bound structures	Secreted products
	Quorum sensing molecules
	Type III secretion: ExoU, ExoT, ExoS, ExoY
	Proteases: elastase, alkaline protease, protease IV
Pili	
	Exotoxin A
Flagellum	
	Pyocyanin
LPS	
	Rhamnolipids
Alginate	
	Phospholipases
	Hemolysins
	Catalase
	Leukocidin

Abbreviations: Exo, exotoxin, LPS, lipopolysaccharide

Pseudomonas Aeruginosa in CF

P. aeruginosa is the leading pathogen causing detrimental chronic lung infections in cystic fibrosis (CF) patients [101, 102]. Cystic fibrosis is the most common genetic disorder in the Caucasian population, which is caused by mutations in the CF transmembrane regulator

(CFTR) gene. In CF patient lungs, the mucociliary escalator is significantly compromised because of a deficient function of CFTR, the major chloride ion channel in airway epithelial cells. The disturbed epithelial ion transport results in severely dehydrated airway secretions and, as a consequence, the thick secretions block the smaller bronchioles and interfere with ciliary beating [103]. It has been proposed that CFTR acts as a specific receptor for *P. aeruginosa* in airway epithelial cells and in its absence bacterial internalization followed by phagocytosis of infected cells does not occur, leading to the major defect in infection clearance [104, 105]. Although precise mechanisms underlying the high susceptibility of CF patients to *P. aeruginosa* are incompletely understood, the significance of host defense in protection against this infection is undoubted. Most CF patients, sooner or later, become infected with *P. aeruginosa*, and the establishment of chronic lung infection caused by this pathogen is the major determinant of morbidity and mortality in CF [106-108].

Mechanisms of Microbial Pathogenesis

Colonization of mucosal surfaces of the respiratory system is the necessary prerequisite for the pulmonary infection caused by *P. aeruginosa*, and this process is initiated upon bacterial adhesion to the host cells (Table 7.6). Interactions of *P. aeruginosa* with airway epithelial cells resulting in adhesion followed by bacterial internalization are mediated by a number of bacterial products including pili, lipopolysaccharide (LPS), flagella, and alginate, and several host cell receptors such as the glycosphingolipid asialoGM1, toll-like receptors (TLR), and CFTR [105, 109-113].

Table 7.6. Pathogenesis of Pseudomonas aeruginosa pulmonary infection in CF patients.

Mechanisms of pathogenesis	Bacterial virulence factors	Mechanisms of host defense	Factors contributing to disease
Colonization of lung epithelial cells Bacterial adhesion	Pili Flagella LPS Alginate	Mucociliary clearance	Abnormal thick mucus Using host molecules as receptors
Progressive spread and growth in the airways	Pili Flagella Pyocyanin	Anti-bacterial peptides and matrilysin in the airway secretions Phagocytosis Complement ROS release by neutrophils	High salt concentrations in airway secretions Presence of cysteine proteases Increased expression of potential receptors Damage to the ciliary function by pyocyanin
Invasion Apoptosis of infected cells	Type III secretion system Pili Flagella LPS ExoA, elastase, alkaline phosphatase, phospholipase C, hemolysins, leukocidin, rhamnolipids Pyocyanin	Desquamation of apoptotic epithelial cells and their phagocytosis	Exposure of potential receptors due to damage of epithelial cells and destruction of tight junctions Apoptosis of neutrophils

Table 7.6 (Continued)

Damage to lung tissue Inflammation	Type III secretion system Pili Flagella LPS ExoA, elastase, alkaline phosphatase, phospholipase C, hemolysins, leukocidin	Activation of signaling from PRR Activation of NF-□B and AP-1 Production of TNF-α and other pro-inflammatory cytokines/chemokines Production of GM-CSF Massive recruitment of neutrophils and their activation	Apoptosis of neutrophils Oxidative stress-mediated damage of lung cells Epithelial cell necrosis
Immune evasion	Quorum sensing system Hypermutations Mutations in *mucA* gene Type III secretion system Lipid A containing palmitate and aminoarabinose	Activation of innate and adaptive immunity	Biofilm formation Alginate overproduction Loss of pili and flagella Loss of antigenic O-side chains of LPS Inhibition of caspase-1 Decreased activation of innate immunity Antibiotic resistance Resistance to antimicrobial peptides Inhibition of surfactant Exhaustion of host defense mechanisms

Abbreviations: AP-1, activator protein A, ExoA, exotoxin A, GM-CSF, granulocyte-macrophage colony-stimulating factor, LPS, lipopolysaccharide, NF-κB, nuclear factor kappa B, PRR, pattern recognition receptor, ROS, reactive oxygen species, TNF, tumor necrosis factor

The surface appendages of *P. aeruginosa*, flagella and pili, which mediate bacterial motility, help to establish close contact with the airway epithelium and facilitate adhesion to the cell surface via binding asialoGM1 [111, 112]. LPS is also capable of binding asialoGM1 [111], as well as the TLR4 complex [114] and CFTR [105]. Integrin receptors, i.e. β1 integrins have been implicated in adherence of *P. aeruginosa* to the airway epithelium, but the structures involved in such interactions are poorly defined [115]. Hence, *P. aeruginosa* use several host molecules as receptors facilitating bacterial adhesion and the following internalization.

Pili are complex surface appendages mediating the twitching motility that allows the bacteria to move across cellular surfaces. Interactions between bacterial pili and epithelial asialoGM1 are critically involved in *P. aeruginosa* adhesion, internalization, and cytotoxicity. *P. aeruginosa* exotoxins, i.e. ExoS, ExoT, ExoU, and ExoY, which are the effectors of the type III secretion systems, are injected into infected cells following pili-mediated bacterial adhesion [116, 117]. Pili are also important for biofilm formation as well as for inflammatory responses induced by *P. aeruginosa* infection of the airway epithelium.

Flagellum is a filamentous polar appendage at the surface of *P. aeruginosa* mediating their swimming motility. Flagella are involved in bacterial adhesion to airway epithelial cells as well as in the resulting inflammatory responses via binding to asialoGM1, along with TLR5 and TLR2 [113]. Flagella are the major pro-inflammatory virulence factors of *P. aeruginosa* and also possess strong immunostimulatory abilities. A structural component of *P. aeruginosa* flagellum, the flagellar cap protein FliD, is directly involved in bacterial adhesion to mucin [118]. This process is important in the process of *P. aeruginosa* colonization of the airways, especially during the early steps of the infectious process.

Lipopolysaccharide (LPS) is the major component of the outer membrane of Gram-negative bacteria; LPS has a hydrophobic domain (lipid A) associated with the hydrophilic tail composed of the core oligosaccharide and the O-specific polysaccharide. The lipid A component of LPS interacts with the TLR4 signaling complex and induces potent pro-inflammatory responses. *P. aeruginosa* are able to synthesize LPS with a variety of penta- and hexa-acylated lipid A structures under different environmental conditions. *P. aeruginosa* causing infections in CF patients synthesize specific lipid A forms containing palmitate and aminoarabinose, and such forms are associated with resistance to cationic antimicrobial peptides and increased inflammatory responses [119]. The variable O-specific polysaccharide chains of *P. aeruginosa* are immunogenic. Environmental strains of *P. aeruginosa* typically express smooth LPS with long O-chains, while strains causing infections in CF patients often lack these O-side chains as well as their antigenic properties. It has been suggested that CFTR expressed on the airway epithelium can bind and extract the outer core oligosaccharide of *P. aeruginosa* LPS and this mediates bacterial internalization and innate immune responses [105]. It was also demonstrated that *P. aeruginosa* LPS is able to bind asialoGM1 expressed on the cornea [111].

Alginate is an extracellular polysaccharide produced by mucoid strains of *P. aeruginosa* isolated from CF patient lung. Alginate is a linear polymer of D-mannuronic acid and L-glucuronic acid; its overproduction is the result of mutations in the anti-sigma factor gene, *mucA*. The production of alginate and flagellin by *P. aeruginosa* is inversely regulated by the alternative sigma factor *AlgT*, acting as a positive regulator of mucoidy and a negative regulator of flagellum-mediated motility [120]. Alginate may be involved in adhesion of *P. aeruginosa* to the airway epithelium via interaction with sialyl-Le(x)-containing glycans found in increased amounts in mucins from CF patients [121]. It was demonstrated that alginate can activate monocytes and macrophages through TLR2 and TLR4 [122], but it is uncertain whether these interactions occur in the airway epithelium.

In normal individuals, the mucociliary clearance successfully prevents adhesion of *P. aeruginosa* to the airway epithelium wiping them away entrapped in mucous. In addition, the bactericidal effect of antibacterial peptides produced by epithelial cells, such as beta-defensins, helps eliminate the pathogens [123]. Any defect in innate immune mechanisms of the respiratory system would predispose to pulmonary infections caused by *P. aeruginosa.* This happens in CF patients, when abnormally thick mucus would interfere with the mucociliary clearance. In these patients, the antimicrobial activity of beta-defensins produced by airway epithelial cells is compromised because of high salt concentrations in airway secretions and presence of cysteine proteases [124, 125]. An increased epithelial expression of potential receptors for *P. aeruginosa* such as asialoGM1, $\alpha 5\beta 1$ integrin, and its ligand fibronectin has been demonstrated [110, 113, 115, 126]. The expression of asialoGM1 is increased on the apical surface of regenerating airway epithelial cells providing additional binding sites to *P. aeruginosa* [127]. In COPD patients, epithelial cell damage caused by cigarette smoke, impaired mucociliary escalator, and mucus hypersecretion can affect clearance of pathogenic microorganisms from the airways and predispose to *P. aeruginosa* bronchial infections. As patients on mechanical lung ventilation experience significant damage to the respiratory epithelium because of endotracheal intubation, they are also at high risk of *P. aeruginosa* pulmonary infections given that this pathogen is common in the hospital environment.

Upon adhesion to epithelial cells, *P. aeruginosa* employ the type III secretion system, which consists of syringe-like apparatus directly injecting bacterial toxins into the cytoplasm of infected cells. The effectors of type III secretion system are exotoxins: ExoY, ExoS, ExoT, ExoU, which have the cytotoxic effect and contribute to the process of bacterial invasion, spreading and dissemination of the pathogen. ExoY is an adenylate cyclase, ExoU is a necrotizing toxin with phospholipase activity, ExoS and ExoT are highly homologous bifunctional proteins with two active domains, i.e. a carboxy-terminal adenosine diphosphate (ADP)-ribosyltransferase (ADPRT) domain and an amino-terminal guanidine triphosphatase (GTPase)-activating (GAP) domain [reviewed by 99]. The pore-forming activity associated with the insertion of the type III secretion system complex can cause cell death as a result of the inflammasome formation and caspase-1 activation. This process is accompanied by secretion of IL-1β and inflammatory responses [128]. The type III secreted toxins are able to alter the distribution of the tight-junction proteins ZO-1 and occludin, and such mechanism allows *P. aeruginosa* transmigrate across polarized airway epithelial cells, contributing to the bacterial invasion [129].

In addition to the effectors of the type III secretion system, the bacteria possess other potent cytotoxins contributing to the microbial pathogenesis. Some secreted bacterial virulence factors, i.e. exotoxin A, elastase, alkaline phosphatase, phospholipase C, hemolysins, and leukocidin promote bacterial invasion as they damage the epithelial cells; this leads to the exposure of potential receptors for bacterial binding [130]. Rhamnolipids, amphiphilic molecules with detergent properties are capable of causing tight-junction alterations destroying the barrier function of the airway epithelium [131]. One of the most potent *P. aeruginosa* toxins is ExoA, which is an ADP-ribosylating enzyme. ExoA is capable of entering eukaryotic cells via receptor-mediated endocytosis and then it catalyses the ADP-ribosylation of eukaryotic elongation factor-2 (eEF-*2*). This leads to the inhibition of protein synthesis and subsequent cellular necrosis [132]. A blue-green pigment pyocyanin disrupts the airway epithelium and impairs the ciliary function [133]. This toxin has oxydoreductive properties; it can oxidase glutathione and inactivate catalase in airway epithelial cells contributing to the oxidative stress-mediated damage of lung tissue [134, 135].

Apoptosis of Host Cells Caused by Pseudomonas Aeruginosa

Programmed cell death, or apoptosis of eukaryotic cells can be induced via two major pathways. The death receptor mediated (extrinsic) pathway is initiated by the aggregation of death receptors, such as Fas (CD95), tumor necrosis factor (TNF) receptor, or tumor necrosis-related *apoptosis*-inducing ligand (TRAIL), upon their binding to specific ligands. Activation of death receptors leads to the recruitment of the adaptor molecule Fas-associated death domain (FADD) followed by recruitment of caspase 8 and its autoproteolytic activation. The mitochondrial (intrinsic) pathway is mediated by cytochrome c release causing the activation of caspase 9. Both pathways converge in the activation of a series of caspases that leads to the cleavage of some intracellular substrates and ultimately results in programmed cell death [136].

P. aeruginosa is able to cause apoptosis of epithelial cells via both extrinsic and intrinsic pathways. Mechanisms of the induction of apoptosis by *P. aeruginosa* infection are complex and involve both cell-associated structures, i.e. pili, flagella, and LPS, and secreted toxic products, such as ExoS, ExoT, ExoU, ExoA, and pyocyanin [112, 137-140] (Table 7.7).

Table 7.7. Apoptosis of epithelial cells induced by Pseudomonas aeruginosa infection.

<table>
<tr><td>Virulence factors</td><td>Cell-associated structures:
Pili
Flagella
LPS</td><td>Secreted toxic products:
T3SS: ExoS, ExoT, ExoU
ExoA
Pyocyanin</td></tr>
<tr><td>Mechanisms of effect</td><td colspan="2">Tageting GTPases Rho, Rac, Cdc-42
Inhibition of ERK1/2 and p38
Activation of JNK
Oxidative stress
Cytochrome c release from mitochondria
Increase in Fas and FasL expression</td></tr>
</table>

Abbreviations: ERK, extracellular signal-regulated kinase, Exo, exotoxin, FasL, Fas ligand, GTPase, guanine-triphospahatase, JNK, c-Jun N-terminal kinase, T3SS, Type 3 secretion system

The products of *P. aeruginosa* type III secretion system are critical in inducing apoptosis. ExoS and ExoT interact with small molecular weight GTPases Rho, Rac, and Cdc-42 of the host cells [141]. Since these GTPases are important in the maintenance of cytoskeletal components and epithelial tight junctions, *P. aeruginosa* toxins acting on these GTP-binding proteins disrupt actin polymerization and dissociate cell-cell contacts important for cell survival [142]. In addition, ExoS can initiate apoptosis in an ADP-ribosyltransferase (ADPRT) activity-dependent manner as it shuts down host cell survival pathways by inhibiting the activation of extracellular signal-regulated kinases (ERK) 1/2 and p38, and triggers pro-apoptotic pathways through activation of c-Jun N-terminal kinase (JNK) 1/2. This results in cytochrome c release from mitochondria followed by the activation of the caspase cascade and cellular apoptosis [117, 139, 143-145]. Via similar mechanism involving the ADPRT domain activity, ExoT also activates the mitochondrial/cytochrome c-dependent apoptotic pathway in epithelial cells [140].

The intrinsic death pathway can be triggered by the oxidative stress as reactive oxygen species (ROS) cause release of cytochrome c from mitochondria and thus initiate the caspase activation cascade [146]. Among *P. aeruginosa* cytotoxins, pyocyanin has oxydoreductive properties and is capable of causing apoptosis of airway epithelial cells and neutrophils via inducing oxidative stress [134, 135].

P. aeruginosa toxins are necessary, but not sufficient for inducing apoptosis, and bacterial-cell contact mediated by pili, flagellin, and LPS is essential for this process [117, 137]. Indeed, the effectors of the type III secretion systems, i.e. ExoS, ExoT, ExoU are injected into infected cells following pili-mediated bacterial adhesion [116, 144]. Moreover, internalization of *P. aeruginosa* is important for apoptosis of infected airway epithelial cells [147]. *P. aeruginosa* infection causes an increase in the expression of Fas and Fas ligand (L) on epithelial cells suggesting the activation of the extrinsic pathway [117, 147, 148]. However, precise molecular mechanisms underlying apoptosis of airway epithelial cells caused by *P. aeruginosa* are still incompletely understood.

Molecular Mechanisms of Inflammation Caused by Pseudomonas Aeruginosa Infection

The hallmark of *P. aeruginosa*-caused disease is the strong inflammatory response of infected tissue characterized by the activation of transcription factors, nuclear factor kappa B (NF-κB), and activator protein (AP)-1. This results in the release of pro-inflammatory mediators, i.e. cytokines tumor necrosis factor (TNF)-α, interleukin (IL)-1, IL-6, chemokines IL-8 and RANTES (regulated on activation, normal T expressed and secreted), increased expression of adhesion molecules (intercellular adhesion molecule, ICAM-1), recruitment of activated neutrophils, and severe tissue damage eventually causing lung failure [130].

Both inflammation and innate immune response to pathogenic microorganisms are initiated following the recognition of pathogen-associated molecular patterns (PAMP) by the pattern recognition receptors (PRR) [149]. Various *P. aeruginosa* products are recognized by toll-like receptors (TLR), the transmembrane glycoproteins that share structural homology and signaling pathways with the IL-1 receptor family [150]. It is established that *P. aeruginosa* flagella are recognized by TLR5 and TLR2, cell wall lipopeptides by TLR2, LPS by TLR4, alginate and ExoS by both TLR2 and TLR4, and bacterial DNA by TLR9 [151, 152]. Upon their activation, TLRs initiate signal transduction leading to the activation of transcription factors, which in turn regulate gene expression of molecules mediating inflammatory responses, and this process eventually leads to the elimination of the pathogen. Recognition of *P. aeruginosa* virulence factors by PRRs involves complex molecular interactions, including communications among different TLRs and other receptors, such as asialoGM1 [150]. Moreover, there is a redundancy in TLR signaling in response to *P. aeruginosa* [152]. *P. aeruginosa* pili bind to the GalNacβ1-4 gal moiety present on asialo-glycolipids and then activate pro-inflammatory signaling pathways via a receptor complex that includes asialoGM1, TLR2, and associated kinases in lipid rafts [126].

During the process of recognition of LPS, lipid A is bound by LPS-binding protein, which transfers LPS to membrane-anchored CD14. The latter molecule facilitates binding of LPS to TLR4 in the presence of a co-factor MD2. This interaction leads to the recruitment of several adaptor molecules containing Toll/IL-1 receptor (TIR) domains, followed by the recruitment of the next series of adaptor molecules, including myeloid differentiation protein MyD88 and TIR-domain-containing adapter-inducing interferon-β (TRIF). The resulting signaling cascade ultimately leads to the activation of gene transcription of pro-inflammatory molecules, such as IL-1, IL-6, IL-8, TNF-α, etc [153].

The recognition of ExoS is even more complex, as it involves both TLR2 and the TLR4/MD2/CD14 complex, and the resulting cellular activation occurs via a MyD88 pathway. Interestingly, the ability to activate TLR2 is localized to the C-terminal domain of ExoS while the TLR4 activity is localized to the N-terminal domain [154].

Activation of airway epithelial cells by *P. aeruginosa* flagella is mediated via asialoGM1, TLR2 and TLR5 interactions [113]. Upon bacterial infection, TLR2 is mobilized into an apical lipid raft receptor complex in epithelial cells, and the association of TLR2 with asialo-glycolipids presented within the context of lipid rafts is essential for the initiation of host cell signaling pathways [126]. Lipid rafts are specialized detergent-insoluble areas of eukaryotic plasma membrane, characterized by high content of cholesterol and sphingolipids as well as of some important signaling molecules along with some typical structural proteins such as caveolins. TLR2 is enriched in caveolin-1-associated lipid raft microdomains present on the apical surface of airway epithelial cells after bacterial infection. These receptor complexes include MyD88, interleukin-1 receptor-activated kinase (IRAK)-1, and TNF receptor-associated factor (TRAF) 6. The signaling capabilities of TLR2 are amplified through its association with the asialo-glycolipids [126]. AsialoGM1 has signaling properties, i.e. activates mitogen activated protein kinase (MAPK) cascade through Ras and Src leading to NF-κB activation followed by pro-inflammatory molecule gene transcription, e.g. IL-8 [155].

P. aeruginosa is also recognized by the IPAF/NLRC4, a member of the Nod-like receptor (NLR) family of intracellular proteins that detect cytosolic PAMPs as well as endogenous danger signals released by cellular damage or stress [156]. The NLR proteins contain a carboxy-terminal leucine-rich repeat (sensor) domain, a central nucleotide-binding NACHT domain, and an amino-terminal protein-protein interaction (signaling) domain. Activation of NLRs results in the assembly of inflammasomes, i.e. multiprotein complexes, which are required for the activation of caspase-1 followed by IL-1β precursor processing into the biologically active IL-1β molecule, ultimately leading to inflammatory responses [167]. Recent studies have established that *P. aeruginosa* activates the IPAF/NLRC4 inflammasome and this process requires flagella and functional type III secretion system [156, 158]. Importantly, flagellin, the structural protein of flagella, which is the dominant pro-inflammatory factor of *P. aeruginosa* can be sensed by both TLR5 and IPAF, but via different regions [158, 159]. Both TLR5 and IPAF recognize the monomeric flagellin rather than the highly polymerized bacterial flagellum [151]. It is possible that monomeric flagellin enters the cytosol of infected cells through the needle complex formed by the type III secretion system apparatus [158]. In addition, following bacterial internalization, acidification of endosomes can release monomeric flagellin from bacteria and allow the recognition by the PRRs [151]. Hence, extremely complex mechanisms mediate epithelial cell responses to flagella. To summarize, flagella activate cells through asialoGM1, TLR2 and TLR5 as well as through intracellular NLR. In addition, NOD1 participates in the innate immune response to *P. aeruginosa* via detection of the peptidoglycan and this process also leads to NF-κB activation [160].

The Role of Airway Epithelial Cells in the Pathogenesis of Pseudomonas Aeruginosa Pulmonary Infections

Airway epithelial cells play critical role in orchestrating both innate defense and inflammatory responses; these cells have been recognized as primary elements generating signals to activate other cells in the lung [161]. In response to *P. aeruginosa* infection, epithelial cells produce various proteins important for host defense, such as anti-bacterial peptides beta-defensins [162], pro-inflammatory cytokines IL-1, IL-6, and TNF-α, chemokines IL-8 and RANTES, the granulocyte-macrophage colony-stimulating factor (GM-CSF) [130], a matrix metalloprotease matrilysin, etc [163].

Using in vivo models of *P. aeruginosa* infection, it has been established that clearance of these bacteria from the lungs is mediated by the activation of NF-κB and resulting TNF-α production [164]. This cytokine is essential for the following activation of both NF-κB and AP-1 transcription factors that induce gene expression of important pro-inflammatory genes.

TNF-α is a critical cytokine in innate immune responses: it regulates recruitment of phagocytic cells and activates them to clear the infection. TNF-α deficient animals succumb to experimental *P. aeruginosa* infection [165, 166]. However, excessive production of TNF-α, for example, during septic shock, is highly detrimental to a host.

IL-1 is a pro-inflammatory cytokine, which exists in two forms, IL-1α and IL-1β, and has been implicated in resistance to *P. aeruginosa* pulmonary infection [167]. The role of IL-1 in *P. aeruginosa* pathogenesis is rather complex and should be seen in the context of a delicate balance between protective innate immune responses and harmful pro-inflammatory activities. Rapid IL-1β release and signaling through the IL-1 receptor have been identified as key steps in the innate immune response, which limits *P. aeruginosa* colonization of the lungs [168]. However, in a model of *P. aeruginosa* pneumonia, reduced production of IL-1 improved host defense against this infection [169]. High levels of IL-1 produced during prolonged time can contribute to the pathogenesis of pulmonary disease caused by *P. aeruginosa* [130].

IL-6 has both pro- and anti-inflammatory characteristics and it is involved in the regulation of adaptive (antibody production, T cell differentiation and activation) and innate (neutrophil activation) immune responses. IL-6 plays essential role in the resolution of acute inflammation and regulates transition from innate to adaptive immunity via the complex interplay between its membrane-bound cognate receptor (IL-6R) and soluble receptor (sIL-6R) [170].

IL-8, a member of the CXC chemokine family, produced by the airway epithelium during the infectious process, is a potent neutrophil attracting chemokine. Several *P. aeruginosa* virulence factors, including pili, flagella, peptidoglycan, and the homoserine lactone autoinducer, are capable of inducing high production of IL-8 by airway epithelial cells [171]. Although neutrophils play a critical role in phagocytosis of *P. aeruginosa* that is indispensable in bacterial clearance, an excessive release of IL-8 by stimulated epithelial cells may contribute to the pathogenesis of *P. aeruginosa* pulmonary infection. Large numbers of activated neutrophils recruited to the lung release toxic products, including reactive oxygen species and proteases. Such mechanisms contribute to the vicious cycle of pulmonary inflammation, in particular, during the chronic *P. aeruginosa* infection in CF patients [172].

GM-CSF is the growth factor produced by airway epithelial cells during *P. aeruginosa* infection. GM-CSF stimulates phagocytic and bactericidal abilities of neutrophils and promotes their survival in the lung [183, 184]. In addition, GM-CSF provides a signal for differentiation of macrophages and dendritic cells and hence stimulates both innate and adaptive immunity [175, 176].

ICAM-1 is an adhesion molecule of the immunoglobulin superfamily expressed by lung epithelial cells under the transcriptional regulation by NF-κB; it is critical in recruitment of inflammatory cells to the infected tissue [177, 178]. ICAM-1 regulates the recruitment of neutrophils and provides a mechanism for their retention in the lung [179].

Hence, *P. aeruginosa* infectious process in the lung is associated with an increased epithelial production of several pro-inflammatory cytokines, chemokines, and adhesion molecules. The resulting cytokine receptor signaling in various cell types is largely responsible for the disease outcome. Although cellular responses initiated by signaling through TNF receptor are critical for the clearance of infection, the effects of IL-1 and IL-8 can be deleterious [130]. In vivo, bacterial injury to the alveolar epithelium can be followed by the release of proinflammatory mediators into the systemic circulation, and the resulting "cytokine storm" is primarily responsible for septic shock frequently associated with acute pneumonia in immunocompromised individuals [180]. During the chronic infectious process, e.g., in CF patient lung, the extensive release of the inflammatory mediators is causing progressive damage to pulmonary tissue perpetuating a vicious cycle of cell injury and inflammation. It is recognized that the lung epithelium is important in both pulmonary innate defense mechanisms and inflammatory responses. However, it is still incompletely understood, how the balance between host defense and inflammation in *P. aeruginosa* pulmonary infection is regulated and what are the crucial mechanisms that determine the outcomes of the infectious process.

The significance of bacterial internalization by airway epithelial cells for the pathogenesis of lung infections caused by *P. aeruginosa* is also controversial. Some studies suggest that internalization of *P. aeruginosa* followed by apoptosis of infected cells is essential for the clearance of bacteria from the lungs, as apoptotic cells are shed and phagocytosed by neutrophils [147, 181]. Apoptosis was found to be necessary for protective host response in a model of acute infection; indeed, Fas- or FasL-deficient mice, but not normal mice, succumbed to sepsis, caused by intranasal infection with *P. aeruginosa* [148].

In CF patient lungs, bacterial invasion would cause apoptosis of infected bronchial and bronchiolar epithelial cells leading to their desquamation and followed by phagocytosis. As bronchial epithelium has several layers of cells, apoptosis of the superficial layer would not significantly compromise the integrity of bronchial wall. This sequence of events can lead either to clearance of infection or to the establishment of chronic infectious process if the host defense mechanisms have been exhausted.

P. aeruginosa internalization can also lead to the release of some pathogen-associated molecules activating innate recognition receptors and consequently would induce innate immune response [151]. In case when apoptotic cells containing bacteria are phagocytosed by alveolar macrophages, this will result in the presentation of bacterial antigens to T cells and initiation of adaptive immune response.

In contrast, in the alveolar epithelial cells, *P. aeruginosa* internalization can potentially protect bacteria from phagocytosis and contribute to their dissemination through the thin layer of the alveolar-capillary membrane. In the acute *P. aeruginosa* pneumonia scenario, the bacteria invade type I alveolar epithelial cells by co-opting the pathway of lipid-raft mediated endocytosis causing tyrosine phosphorylation of the key protein of lipid rafts, caveolin-2 [182]. This pathway can provide bacteria some protection from clearance by alveolar macrophages in the intracellular environment. As bacteria may leave the infected cells via exocytosis from the basolateral site and then penetrate the alveolar-capillary barrier, this pathway would lead to *P. aeruginosa* dissemination within the host and to systemic toxic effect caused by the release of bacterial virulence factors. A single-layer, thin alveolar epithelium undergoing apoptosis as a result of bacterial invasion, would not be able to prevent bacteremia [182]. Dissemination of *P. aeruginosa* from the lungs to the systemic circulation resulting in bacteremia may occur during acute infectious process, such as ventilator-associated pneumonia. However, this never happens during the chronic pulmonary infections in CF patients. Hence, the outcomes of bacterial internalization into airway epithelial cells depend on the particular compartment of the airways, which is infected, as well as on the presence of specific bacterial virulence factors.

Immune Evasion

P. aeruginosa possess remarkable abilities of evading immune mechanisms via disturbing normal lung physiology and host defense. The ADPRT activity of the type 3 secretion system effector protein ExoS inhibits the caspase-1 mediated maturation and secretion of IL-1β [183]. In addition, ExoU is able to inhibit caspase-1 production. Therefore *P. aeruginosa* strains expressing the type 3 secretion system can evade the immune system [156]. Indeed, it was demonstrated that *P. aeruginosa* strains that secrete ExoU cause more severe disease [184].

In the process of CF lung disease, *P. aeruginosa* undergo significant genetic and phenotypic changes, i.e. lose motility, decrease production of some virulence factors, such as pyocyanin and proteases, change LPS structure, and convert to a mucoid phenotype associated with large production of the exopolysaccharide alginate [195, 196]. Such genetic adaptations of *P. aeruginosa* to the CF lung environment are mediated by bacterial hypermutations [187]. A transition from a non-mucoid to a mucoid, alginate-hyperproducing phenotype is caused by mutations in the *mucA* gene [198]. Such changes may help bacteria to evade the host defense mechanisms and contribute to the establishment of chronic *P. aeruginosa* infection [185, 186]. Overexpression of alginate protects bacteria from phagocytosis, reactive oxygen species produced by neutrophils, antibiotics, and interferes with host response, i.e. mucociliary clearance [189-191].

P. aeruginosa target pulmonary surfactant, a surface-active mixture of phospholipids and proteins important in host defense. *P. aeruginosa* proteases, i.e. elastase degrade surfactant protein A (SP-A) and SP-D [192]. In addition, chronic *P. aeruginosa* lung infection reduces surfactant levels by inhibiting its biosynthesis [193]. Alginate contributes to transcriptional repression of phosphocholine cytidyltransferase, the enzyme required for synthesis of the major surfactant phospholipid [193]. As decreased levels of surfactant promote airway collapse, the resulting small airway obstruction interferes with clearance of pathogens. Global analysis of epithelial cell gene expression in response to motile (flagellin-expressing) and mucoid (alginate-expressing) strains has demonstrated that the production of alginate attenuates the magnitude of the host response to *P. aeruginosa* [194]. In vitro, alginate-overproducing bacteria did not cause an increased expression of matrilysin, a matrix metalloprotease involved in host response, or human beta-defensin-2, although flagellin-expressing strain did [194].

P. aeruginosa possess the quorum-sensing systems serving as a mechanism to coordinate expression of genes important for adaptation to the environment through the production of small diffusible molecules called autoinducers. *P. aeruginosa* produce acyl-homoserine lactones acting as quorum-sensing molecules, which bind LasR/RhlR transcriptional activators and then induce coordinated gene expression in microbial communities [205]. The activation of the quorum-sensing systems promotes the formation of biofilms that is an important characteristic of *P. aeruginosa* growth in CF patient lungs. The biofilm type of growth contributes to the persistence of pulmonary infections in CF because bacteria growing in biofilms are more resistant to host defense mechanisms and antibiotics compared to bacteria growing in the planktonic form. In particular, biofilm formation during *P. aeruginosa* pulmonary infection inhibits neutrophil function [179]. In addition, acyl-homoserine lactones are able to enter host cells and interfere with intracellular signaling pathways affecting normal cellular responses, i.e. induce apoptosis, as well as pro-inflammatory cytokine or COX-2 release [196, 197]. Via this pathway, *P. aeruginosa* can exploit host cell defense mechanisms. *P. aeruginosa* homoserine lactone C12 selectively impairs the regulation of NF-κB functions in activated eukaryotic cells. The consequence is specific repression of stimulus-mediated induction of NF-κB-responsive genes encoding inflammatory cytokines and other immune regulators [17]. This strategy attenuating the innate immune system helps bacteria establish and maintain local persistent infection, i.e. in CF patients.

In the process of *P. aeruginosa* chronic infection in CF patients, changes to LPS structure occur. Bacteria lose antigenic O-specific polysaccharide chains and this affects their ability to elicit adaptive immune responses. *P. aeruginosa* causing infections in CF patients synthesize specific lipid A forms containing palmitate and aminoarabinose, and such forms are associated with resistance to cationic antimicrobial peptides and increased inflammatory responses [119]. In the course of CF pulmonary disease, bacteria also lose pili and flagella critical for the initial phase of infection; these structures function as ligands for phagocytic cells and stimulate the recruitment of neutrophils [198, 199]. The mutant bacteria, which lost the expression of flagella and pili are less immunogenic, therefore they are selected and persist in the CF patient lung.

P. aeruginosa strains causing early stages of the infection in CF patients are able to elicit strong innate and adaptive immune responses, which help the host temporarily eradicate the infection during the period of intermittent colonization of the lung. However, with time, the loss of bacterial virulence factors would result in diminished innate immune responses to *P. aeruginosa.* Such changes can provide survival advantages to the bacteria, which then successfully establish the chronic infection. During the chronic infection, *P. aeruginosa* retain high pro-inflammatory abilities that contribute to the pathogenesis of CF lung disease [200]. Continuing stimulation of the host cells by pro-inflammatory mediators along with escalating tissue damage would eventually lead to exhaustion of host defense mechanisms, and along with developing resistance to many antibiotics would make the eradication of the infection impossible.

Conclusion

Pathogenic microorganisms can successfully overcome the host defense mechanisms in the respiratory system, e.g. using host cellular receptors to adhere to and invade the airway epithelium. The pathogens produce potent virulence factors, which destroy normal tissue barriers and disable immune defenses. For the outcomes of the infection, host-pathogen interactions are critically important. *S. pneumoniae* can often colonize the upper respiratory tract of healthy individuals, but it never causes lower airway infection unless there is a breach in immune defense, e.g. an impaired mucociliary clearance, lack of opsonizing antibody, or defect in innate immune activation. This pathogen uses numerous strategies to destroy the tissue barriers, such as production of pneumolysin, which makes pores in cellular membranes, or exploitation of the plasminogen-plasmin system helping bacteria to penetrate the epithelial basement membrane. However, in an immune host, *S. pneumoniae* does not have a chance to use its potent virulence factors because anti-capsular polysaccharide antibodies opsonize bacteria and phagocytic leukocytes rapidly destroy them. In the case of *P. aeruginosa*, these bacteria are efficiently eliminated by the mucociliary clearance and can only cause pulmonary infections if this defense mechanism is impaired. This happens in CF patient lung, when severe dehydration of airway secretions affects the mucociliary function. When *P. aeruginosa* colonize the lung, they use remarkable strategies to evade the immune defense and consequently cannot be eradicated. Moreover, *P. aeruginosa* undergo the microevolution in the CF patient lung environment, i.e. lose immunostimulatory virulence factors, develop the mucoid phenotype and antibiotic resistance, and grow in biofilms that protect against the immune system. These examples show the amazing capabilities of pathogenic microorganisms to overcome the host defense mechanisms. This explains why the pulmonary pathogens are so powerful and still cause significant morbidity and mortality despite all the achievements of modern medicine, including antibiotics, vaccines, and hygiene.

References

[1] Guyton, AC; Hall, JE. *Textbook of medical physiology.* 11th ed. Philadelphia PA: Elsevier Saunders; 2006.

[2] Bowdish, DM; Davidson, DJ; Hancock, RE. Immunomodulatory properties of defensins and cathelicidins. *Current Topics in Microbiology and Immunology,* 2006 306, 27-66.

[3] Ip, WK; Takahashi, K; Ezekowitz, RA; Stuart, LM. Mannose-binding lectin and innate immunity. *Immunological Reviews,* 2009 230, 9-21.

[4] Waters, P; Vaid, M; Kishore, U; Madan, T. Lung surfactant proteins A and D as pattern recognition proteins. *Advances in Experimental Medicine and Biology,* 2009 653, 74-97.

[5] Mantovani, A; Garlanda, C; Doni, A; Bottazzi, B. Pentraxins in innate immunity: from C-reactive protein to the long pentraxin PTX3. *Journal of Clinical Immunology,* 2008 28, 1-13.

[6] Carroll, MC. The complement system in regulation of adaptive immunity. *Nature Immunology,* 2004 5, 981-986.

[7] Martin, TR; Frevert, CW. Innate immunity in the lungs. *Proceedings of the American Thoracic Society,* 2005 2, 403-411.

[8] Schleimer, RP; Kato, A; Kern, R; Kuperman, D; Avila, PC. Epithelium: at the interface of innate and adaptive immune responses. *Journal of Allergy and Clinical Immunology,* 2007 120, 1279-1284.

[9] Palm, NW; Medzhitov, R. Pattern recognition receptors and control of adaptive immunity. *Immunological Reviews,* 2009 227, 221-233.

[10] Ku, CL; von Bernuth, H; Picard, C; Zhang, SY; Chang, HH; Yang, K; et al. Selective predisposition to bacterial infections in IRAK-4-deficient children: IRAK-4-dependent TLRs are otherwise redundant in protective immunity. *The Journal of Experimental Medicine,* 2007 204, 2407-2422.

[11] von Bernuth, H; Picard, C; Jin, Z; Pankla, R; Xiao, H; Ku, CL; et al. Pyogenic bacterial infections in humans with MyD88 deficiency. *Science,* 2008 321, 691-696.

[12] Wong, KW; Isberg, RR. Emerging views on integrin signaling via Rac1 during invasin-promoted bacterial uptake. *Current Opinion in Microbiology,* 2005 8, 4-9.

[13] Merz, AJ; So, M. Interactions of pathogenic neisseriae with epithelial cell membranes. *Annual Review of Cell and Developmental Biology,* 2000 16, 423-457.

[14] Ishibashi, Y; Claus, S; Relman, DA. Bordetella pertussis filamentous hemagglutinin interacts with a leukocyte signal transduction complex and stimulates bacterial adherence to monocyte CR3 (CD11b/CD18). *The Journal of Experimental Medicine,* 1994 180, 1225-1233.

[15] Kraus, D; Peschel, A. Staphylococcus aureus evasion of innate antimicrobial defense. *Future Microbiology,* 2008 3, 437-451.

[16] Flynn, JL; Chan, J. Immune evasion by Mycobacterium tuberculosis: living with the enemy. *Current Opinion in Immunology,* 2003 15, 450-455.

[17] Kravchenko, VV; Kaufmann, GF; Mathison, JC; Scott, DA; Katz, AZ; Grauer, DC; et al. Modulation of gene expression via disruption of NF-kappaB signaling by a bacterial small molecule. *Science,* 2008 321, 259-263.

[18] Kunert, A; Losse, J; Gruszin, C; Huhn, M; Kaendler, K; Mikkat, S; et al. Immune evasion of the human pathogen Pseudomonas aeruginosa: elongation factor Tuf is a factor H and plasminogen binding protein. *The Journal of Immunology,* 2007 179, 2979-2988.

[19] Husain, AN. The lung. In: Kumar, A; Fausto, A. Robbins and Cotran pathologic basis of disease. 8th ed. City State: Elsevier Saunders; 2010. p. 677-738.

[20] Gu, J; Korteweg, C. Pathology and pathogenesis of severe acute respiratory syndrome. *American Journal of Pathology,* 2007 170, 1136-1147.

[21] Adderson, EE. Antibody repertoires in infants and adults: effects of T-independent and T-dependent immunizations. *Springer Seminars in Immunopathology,* 2001 23, 387-403.

[22] Finn, A; Jenkinson, HF. The pneumococcus: 'old man's friend' and children's foe. *Microbiology,* 2006 152, 281-283.

[23] Kadioglu, A; Weiser, JN; Paton, JC; Andrew, PW. The role of Streptococcus pneumoniae virulence factors in host respiratory colonization and disease. *Nature Reviews Microbiology,* 2008 6, 288-301.

[24] Whitney, CG; Farley, MM; Hadler, J; Harrison, LH; Bennett, NM; Lynfield, R; et al. Decline in invasive pneumococcal disease after the introduction of protein-polysaccharide conjugate vaccine. *The New England Journal of Medicine,* 2003 348, 1737-1746.

[25] Bogaert, D; De Groot, R; Hermans, PW. Streptococcus pneumoniae colonisation: the key to pneumococcal disease. *The Lancet Infectious Diseases,* 2004 4, 144-154.

[26] Cundell, DR; Gerard, NP; Gerard, C; Idanpaan-Heikkila, I; Tuomanen, EI. Streptococcus pneumoniae anchor to activated human cells by the receptor for platelet-activating factor. *Nature,* 1995 377, 435-438.

[27] Ring, A; Weiser, JN; Tuomanen, EI. Pneumococcal trafficking across the blood-brain barrier. Molecular analysis of a novel bidirectional pathway. *The Journal of Clinical Investigation,* 1998 102, 347-360.

[28] Vollmer, W; Tomasz, A. Identification of the teichoic acid phosphorylcholine esterase in Streptococcus pneumoniae. *Molecular Microbiology,* 2001 39, 1610-1622.

[29] Hermoso, JA; Lagartera, L; Gonzalez, A; Stelter, M; Garcia, P; Martinez-Ripoll, M; et al. Insights into pneumococcal pathogenesis from the crystal structure of the modular

teichoic acid phosphorylcholine esterase Pce. *Nature Structural & Molecular Biology,* 2005 12, 533-538.

[30] Anderton, JM; Rajam, G; Romero-Steiner, S; Summer, S; Kowalczyk, AP; Carlone, GM; et al. E-cadherin is a receptor for the common protein pneumococcal surface adhesin A (PsaA) of Streptococcus pneumoniae. *Microbial Pathogenesis,* 2007 42, 225-236.

[31] Steinberg, MS; McNutt, PM. Cadherins and their connections: adhesion junctions have broader functions. *Current Opinion in Cell Biology,* 1999 11, 554-560.

[32] Rajam, G; Phillips, DJ; White, E; Anderton, J; Hooper, CW; Sampson, JS; et al. A functional epitope of the pneumococcal surface adhesin A activates nasopharyngeal cells and increases bacterial internalization. *Microbial Pathogenesis,* 2008 44, 186-196.

[33] Barocchi, MA; Ries, J; Zogaj, X; Hemsley, C; Albiger, B; Kanth, A; et al. A pneumococcal pilus influences virulence and host inflammatory responses. *Proceedings of the National Academy of Sciences,* 2006 103, 2857-2862.

[34] Hermans, PW; Adrian, PV; Albert, C; Estevao, S; Hoogenboezem, T; Luijendijk, IH; et al. The streptococcal lipoprotein rotamase A (SlrA) is a functional peptidyl-prolyl isomerase involved in pneumococcal colonization. *The Journal of Biological Chemistry,* 2006 281, 968-976.

[35] Rennemeier, C; Hammerschmidt, S; Niemann, S; Inamura, S; Zahringer, U; Kehrel, BE. Thrombospondin-1 promotes cellular adherence of Gram-positive pathogens via recognition of peptidoglycan. *The Journal of the Federation of American Societies for Experimental Biology,* 2007 21, 3118-3132.

[36] Bergmann, S; Lang, A; Rohde, M; Agarwal, V; Rennemeier, C; Grashoff, C; et al. Integrin-linked kinase is required for vitronectin-mediated internalization of Streptococcus pneumoniae by host cells. *Journal of Cell Science,* 2009 122, 256-267.

[37] Ulanova, M; Gravelle, S; Barnes, R. The role of epithelial integrin receptors in recognition of pulmonary pathogens. *Journal of Innate Immunity*, 2008 1, 1-14.

[38] Pracht, D; Elm, C; Gerber, J; Bergmann, S; Rohde, M; Seiler, M; et al. PavA of Streptococcus pneumoniae modulates adherence, invasion, and meningeal inflammation. *Infection and Immunity,* 2005 73, 2680-2689.

[39] Holmes, AR; McNab, R; Millsap, KW; Rohde, M; Hammerschmidt, S; Mawdsley, JL; et al. The pavA gene of Streptococcus pneumoniae encodes a fibronectin-binding protein that is essential for virulence. *Molecular Microbiology,* 2001 41, 1395-1408.

[40] Shivshankar, P; Sanchez, C; Rose, LF; Orihuela, CJ. The Streptococcus pneumoniae adhesin PsrP binds to Keratin 10 on lung cells. *Molecular Microbiology,* 2009 73, 663-679.

[41] Zhang, JR; Mostov, KE; Lamm, ME; Nanno, M; Shimida, S; Ohwaki, M; et al. The polymeric immunoglobulin receptor translocates pneumococci across human nasopharyngeal epithelial cells. *Cell,* 2000 102, 827-837.

[42] Lu, L; Lamm, ME; Li, H; Corthesy, B; Zhang, JR. The human polymeric immunoglobulin receptor binds to Streptococcus pneumoniae via domains 3 and 4. *The Journal of Biological Chemistry,* 2003 278, 48178-48187.

[43] Hammerschmidt, S; Tillig, MP; Wolff, S; Vaerman, JP; Chhatwal, GS. Species-specific binding of human secretory component to SpsA protein of Streptococcus pneumoniae via a hexapeptide motif. *Molecular Microbiology,* 2000 36, 726-736.

[44] Agarwal, V; Hammerschmidt, S. Cdc42 and the phosphatidylinositol 3-kinase-Akt pathway are essential for PspC-mediated internalization of pneumococci by respiratory epithelial cells. *The Journal of Biological Chemistry,* 2009 284, 19427-19436.

[45] Bergmann, S; Rohde, M; Chhatwal, GS; Hammerschmidt, S. alpha-Enolase of Streptococcus pneumoniae is a plasmin(ogen)-binding protein displayed on the bacterial cell surface. *Molecular Microbiology,* 2001 40, 1273-1287.

[46] Bergmann, S; Rohde, M; Hammerschmidt, S. Glyceraldehyde-3-phosphate dehydrogenase of Streptococcus pneumoniae is a surface-displayed plasminogen-binding protein. *Infection and Immunity,* 2004 72, 2416-2419.

[47] Attali, C; Durmort, C; Vernet, T; Di Guilmi, AM. The interaction of Streptococcus pneumoniae with plasmin mediates transmigration across endothelial and epithelial monolayers by intercellular junction cleavage. *Infection and Immunity,* 2008 76, 5350-5356.

[48] Ehinger, S; Schubert, WD; Bergmann, S; Hammerschmidt, S; Heinz, DW. Plasmin(ogen)-binding alpha-enolase from Streptococcus pneumoniae: crystal structure and evaluation of plasmin(ogen)-binding sites. *Journal of Molecular Biology,* 2004 343, 997-1005.

[49] Attali, C; Frolet, C; Durmort, C; Offant, J; Vernet, T; Di Guilmi, AM. Streptococcus pneumoniae choline-binding protein E interaction with plasminogen/plasmin stimulates migration across the extracellular matrix. *Infection and Immunity,* 2008 76, 466-476.

[50] Tilley, SJ; Orlova, EV; Gilbert, RJ; Andrew, PW; Saibil, HR. Structural basis of pore formation by the bacterial toxin pneumolysin. *Cell,* 2005 121, 247-256.

[51] Iliev, AI; Djannatian, JR; Nau, R; Mitchell, TJ; Wouters, FS. Cholesterol-dependent actin remodeling via RhoA and Rac1 activation by the Streptococcus pneumoniae toxin pneumolysin. *Proceedings of the National Academy of Sciences,* 2007 104, 2897-2902.

[52] Finlay, BB. Bacterial virulence strategies that utilize Rho GTPases. *Current Topics in Microbiology and Immunology,* 2005 291, 1-10.

[53] Witzenrath, M; Gutbier, B; Hocke, AC; Schmeck, B; Hippenstiel, S; Berger, K; et al. Role of pneumolysin for the development of acute lung injury in pneumococcal pneumonia. *Critical Care Medicine,* 2006 34, 1947-1954.

[54] Weber, JR; Freyer, D; Alexander, C; Schroder, NW; Reiss, A; Kuster, C; et al. Recognition of pneumococcal peptidoglycan: an expanded, pivotal role for LPS binding protein. *Immunity,* 2003 19, 269-279.

[55] Opitz, B; Puschel, A; Schmeck, B; Hocke, AC; Rosseau, S; Hammerschmidt, S; et al. Nucleotide-binding oligomerization domain proteins are innate immune receptors for internalized Streptococcus pneumoniae. *The Journal of Biological Chemistry,* 2004 279, 36426-36432.

[56] van Rossum, AM; Lysenko, ES; Weiser, JN. Host and bacterial factors contributing to the clearance of colonization by Streptococcus pneumoniae in a murine model. *Infection and Immunity,* 2005 73, 7718-7726.

[57] Malley, R; Henneke, P; Morse, SC; Cieslewicz, MJ; Lipsitch, M; Thompson, CM; et al. Recognition of pneumolysin by Toll-like receptor 4 confers resistance to pneumococcal infection. *Proceedings of the National Academy of Sciences,* 2003 100, 1966-1971.

[58] Albiger, B; Dahlberg, S; Sandgren, A; Wartha, F; Beiter, K; Katsuragi, H; et al. Toll-like receptor 9 acts at an early stage in host defence against pneumococcal infection. *Cellular Microbiology,* 2007 9, 633-644.

[59] Vos, Q; Lees, A; Wu, ZQ; Snapper, CM; Mond, JJ. B-cell activation by T-cell-independent type 2 antigens as an integral part of the humoral immune response to pathogenic microorganisms. *Immunological Reviews,* 2000 176, 154-170.

[60] Jeurissen, A; Wuyts, M; Kasran, A; Ramdien-Murli, S; Boon, L; Ceuppens, JL; et al. Essential role for CD40 ligand interactions in T lymphocyte-mediated modulation of the murine immune response to pneumococcal capsular polysaccharides. *The Journal of Immunology,* 2002 168, 2773-2781.

[61] Kawakami, K; Yamamoto, N; Kinjo, Y; Miyagi, K; Nakasone, C; Uezu, K; et al. Critical role of Valpha14+ natural killer T cells in the innate phase of host protection against Streptococcus pneumoniae infection. *European Journal of Immunology,* 2003 33, 3322-3330.

[62] McCool, TL; Cate, TR; Moy, G; Weiser, JN. The immune response to pneumococcal proteins during experimental human carriage. *The Journal of Experimental Medicine,* 2002 195, 359-365.

[63] Kadioglu, A; Coward, W; Colston, MJ; Hewitt, CR; Andrew, PW. CD4-T-lymphocyte interactions with pneumolysin and pneumococci suggest a crucial protective role in the host response to pneumococcal infection. *Infection and Immunity,* 2004 72, 2689-2697.

[64] Zhang, Q; Bernatoniene, J; Bagrade, L; Paton, JC; Mitchell, TJ; Hammerschmidt, S; et al. Regulation of production of mucosal antibody to pneumococcal protein antigens by T-cell-derived gamma interferon and interleukin-10 in children. *Infection and Immunity,* 2006 74, 4735-4743.

[65] Kirby, AC; Coles, MC; Kaye, PM. Alveolar macrophages transport pathogens to lung draining lymph nodes. *The Journal of Immunology,* 2009 183, 1983-1989.

[66] Casal, J; Tarrago, D. Immunity to Streptococcus pneumoniae: Factors affecting production and efficacy. *Current Opinion in Infectious Diseases,* 2003 16, 219-224.

[67] Suresh, MV; Singh, SK; Ferguson, DA,Jr; Agrawal, A. Role of the property of C-reactive protein to activate the classical pathway of complement in protecting mice from pneumococcal infection. *The Journal of Immunology,* 2006 176, 4369-4374.

[68] Gould, JM; Weiser, JN. The inhibitory effect of C-reactive protein on bacterial phosphorylcholine platelet-activating factor receptor-mediated adherence is blocked by surfactant. *The Journal of Infectious Diseases,* 2002 186, 361-371.

[69] Kerr, AR; Irvine, JJ; Search, JJ; Gingles, NA; Kadioglu, A; Andrew, PW; et al. Role of inflammatory mediators in resistance and susceptibility to pneumococcal infection. *Infection and Immunity,* 2002 70, 1547-1557.

[70] Brown, JS; Hussell, T; Gilliland, SM; Holden, DW; Paton, JC; Ehrenstein, MR; et al. The classical pathway is the dominant complement pathway required for innate immunity to Streptococcus pneumoniae infection in mice. *Proceedings of the National Academy of Sciences,* 2002 99, 16969-16974.

[71] Sato, S; Ouellet, N; Pelletier, I; Simard, M; Rancourt, A; Bergeron, MG. Role of galectin-3 as an adhesion molecule for neutrophil extravasation during streptococcal pneumonia. *The Journal of Immunology,* 2002 168, 1813-1822.

[72] Gauthier, JF; Fortin, A; Bergeron, Y; Dumas, MC; Champagne, ME; Bergeron, MG. Differential contribution of bacterial N-formyl-methionyl-leucyl- phenylalanine and host-derived CXC chemokines to neutrophil infiltration into pulmonary alveoli during murine pneumococcal pneumonia. *Infection and Immunity,* 2007 75, 5361-5367.

[73] Beisswenger, C; Coyne, CB; Shchepetov, M; Weiser, JN. Role of p38 MAP kinase and transforming growth factor-beta signaling in transepithelial migration of invasive bacterial pathogens. *The Journal of Biological Chemistry,* 2007 282, 28700-28708.

[74] Noske, N; Kammerer, U; Rohde, M; Hammerschmidt, S. Pneumococcal interaction with human dendritic cells: phagocytosis, survival, and induced adaptive immune response are manipulated by PavA. *The Journal of Immunology,* 2009 183, 1952-1963.

[75] Nelson, AL; Roche, AM; Gould, JM; Chim, K; Ratner, AJ; Weiser, JN. Capsule enhances pneumococcal colonization by limiting mucus-mediated clearance. *Infection and Immunity,* 2007 75, 83-90.

[76] Weiser, JN; Bae, D; Epino, H; Gordon, SB; Kapoor, M; Zenewicz, LA; et al. Changes in availability of oxygen accentuate differences in capsular polysaccharide expression by phenotypic variants and clinical isolates of Streptococcus pneumoniae. *Infection and Immunity,* 2001 69, 5430-5439.

[77] Kim, JO; Weiser, JN. Association of intrastrain phase variation in quantity of capsular polysaccharide and teichoic acid with the virulence of Streptococcus pneumoniae. *The Journal of Infectious Diseases,* 1998 177, 368-377.

[78] Ren, B; Szalai, AJ; Thomas, O; Hollingshead, SK; Briles, DE. Both family 1 and family 2 PspA proteins can inhibit complement deposition and confer virulence to a capsular serotype 3 strain of Streptococcus pneumoniae. *Infection and Immunity,* 2003 71, 75-85.

[79] Shaper, M; Hollingshead, SK; Benjamin, WH,Jr; Briles, DE. PspA protects Streptococcus pneumoniae from killing by apolactoferrin, and antibody to PspA enhances killing of pneumococci by apolactoferrin [corrected]. *Infection and Immunity,* 2004 72, 5031-5040.

[80] Dave, S; Carmicle, S; Hammerschmidt, S; Pangburn, MK; McDaniel, LS. Dual roles of PspC, a surface protein of Streptococcus pneumoniae, in binding human secretory IgA and factor H. *The Journal of Immunology,* 2004 173, 471-477.

[81] Hammerschmidt, S; Agarwal, V; Kunert, A; Haelbich, S; Skerka, C; Zipfel, PF. The host immune regulator factor H interacts via two contact sites with the PspC protein of Streptococcus pneumoniae and mediates adhesion to host epithelial cells. *The Journal of Immunology,* 2007 178, 5848-5858.

[82] Zipfel, PF; Hallstrom, T; Hammerschmidt, S; Skerka, C. The complement fitness factor H: role in human diseases and for immune escape of pathogens, like pneumococci. *Vaccine,* 2008 26 Suppl 8, I67-74.

[83] Jarva, H; Hellwage, J; Jokiranta, TS; Lehtinen, MJ; Zipfel, PF; Meri, S. The group B streptococcal beta and pneumococcal Hic proteins are structurally related immune evasion molecules that bind the complement inhibitor factor H in an analogous fashion. *The Journal of Immunology,* 2004 172, 3111-3118.

[84] Jarva, H; Janulczyk, R; Hellwage, J; Zipfel, PF; Bjorck, L; Meri, S. Streptococcus pneumoniae evades complement attack and opsonophagocytosis by expressing the pspC locus-encoded Hic protein that binds to short consensus repeats 8-11 of factor H. *The Journal of Immunology,* 2002 168, 1886-1894.

[85] Batten, MR; Senior, BW; Kilian, M; Woof, JM. Amino acid sequence requirements in the hinge of human immunoglobulin A1 (IgA1) for cleavage by streptococcal IgA1 proteases. *Infection and Immunity,* 2003 71, 1462-1469.

[86] Kilian, M; Reinholdt, J; Lomholt, H; Poulsen, K; Frandsen, EV. Biological significance of IgA1 proteases in bacterial colonization and pathogenesis: critical evaluation of experimental evidence. *Acta Pathologica, Microbiologica et Immunologica Scandinavica,* 1996 104, 321-338.

[87] Weiser, JN; Bae, D; Fasching, C; Scamurra, RW; Ratner, AJ; Janoff, EN. Antibody-enhanced pneumococcal adherence requires IgA1 protease. *Proceedings of the National Academy of Sciences,* 2003 100, 4215-4220.

[88] Dockrell, DH; Marriott, HM; Prince, LR; Ridger, VC; Ince, PG; Hellewell, PG; et al. Alveolar macrophage apoptosis contributes to pneumococcal clearance in a resolving model of pulmonary infection. *The Journal of Immunology,* 2003 171, 5380-5388.

[89] Kirby, AC; Newton, DJ; Carding, SR; Kaye, PM. Pulmonary dendritic cells and alveolar macrophages are regulated by gammadelta T cells during the resolution of S. pneumoniae-induced inflammation. *The Journal of Pathology,* 2007 212, 29-37.

[90] Noone, PG; Leigh, MW; Sannuti, A; Minnix, SL; Carson, JL; Hazucha, M; et al. Primary ciliary dyskinesia: diagnostic and phenotypic features. *American Journal of Respiratory and Critical Care Medicine*, 2004 169, 459-467.

[91] Pollard, AJ; Heale, JP; Tsang, A; Massing, B; Speert, DP. Nonopsonic phagocytosis of Pseudomonas aeruginoas: insights from an infant with leukocyte adhesion deficiency. *The Pediatric Infectious Disease Journal*, 2001 20, 452-454.

[92] Hakki, M; Limaye, AP; Kim, HW; Kirby, KA; Corey, L; Boeckh, M. Invasive Pseudomonas aeruginosa infections: high rate of recurrence and mortality after hematopoietic cell transplantation. *Bone Marrow Transplantation*, 2007 39, 687-693.

[93] Franzetti, F; Grassini, A; Piazza, M; Degl'innocenti, M; Bandera, A; Gazzola, L; et al. Nosocomial bacterial pneumonia in HIV-infected patients: risk factors for adverse outcome and implications for rational empiric antibiotic therapy. *Infection*, 2006 34, 9-16.

[94] Ramphal, R. Changes in the etiology of bacteremia in febrile neutropenic patients and the susceptibilities of the currently isolated pathogens. *Clinical Infectious Diseases*, 2004 39, S25-31.

[95] Berghmans, T; Sculier, JP; Klastersky, J. A prospective study of infections in lung cancer patients admitted to the hospital. *Chest*, 2003 124, 114-120.

[96] Crouch Brewer, S; Wunderink, RG; Jones, CB; Leeper, KV,Jr. Ventilator-associated pneumonia due to Pseudomonas aeruginosa. *Chest*, 1996 109, 1019-1029.

[97] Garau, J; Gomez, L. Pseudomonas aeruginosa pneumonia. *Current Opinion in Infectious Diseases*, 2003 16, 135-143.

[98] Lieberman, D; Lieberman, D. Pseudomonal infections in patients with COPD: epidemiology and management. *American Journal of Respiratory Medicine*, 2003 2, 459-468.

[99] Kipnis, E; Sawa, T; Wiener-Kronish, J. Targeting mechanisms of Pseudomonas aeruginosa pathogenesis. *Médecine et Maladies Infectieuses*, 2006 36, 78-91.

[100] Poole, K. Aminoglycoside resistance in Pseudomonas aeruginosa. *Antimicrobial Agents and Chemotherapy*, 2005 49, 479-487.

[101] Hart, CA; Winstanley, C. Persistent and aggressive bacteria in the lungs of cystic fibrosis children. *British Medical Bulletin*, 2002 61, 81-96.

[102] Saiman, L; Siegel, J. Infection control in cystic fibrosis. *Clinical Microbiology Reviews*, 2004 17, 57-71.

[103] Matsui, H; Grubb, BR; Tarran, R; Randell, SH; Gatzy, JT; Davis, CW; et al. Evidence for periciliary liquid layer depletion, not abnormal ion composition, in the pathogenesis of cystic fibrosis airways disease. *Cell*, 1998 95, 1005-1015.

[104] Pier, GB; Grout, M; Zaidi, TS. Cystic fibrosis transmembrane conductance regulator is an epithelial cell receptor for clearance of Pseudomonas aeruginosa from the lung. *Proceedings of the National Academy of Sciences*, 1997 94, 12088-12093.

[105] Schroeder, TH; Lee, MM; Yacono, PW; Cannon, CL; Gerceker, AA; Golan, DE; et al. CFTR is a pattern recognition molecule that extracts Pseudomonas aeruginosa LPS from the outer membrane into epithelial cells and activates NF-kappa B translocation. *Proceedings of the National Academy of Sciences*, 2002 99, 6907-6912.

[106] Henry, RL; Mellis, CM; Petrovic, L. Mucoid Pseudomonas aeruginosa is a marker of poor survival in cystic fibrosis. *Pediatric Pulmonology*, 1992 12, 158-161.

[107] Nixon, GM; Armstrong, DS; Carzino, R; Carlin, JB; Olinsky, A; Robertson, CF; et al. Clinical outcome after early Pseudomonas aeruginosa infection in cystic fibrosis. *Journal of Pediatrics*, 2001 138, 699-704.

[108] Burns, JL; Gibson, RL; McNamara, S; Yim, D; Emerson, J; Rosenfeld, M; et al. Longitudinal assessment of Pseudomonas aeruginosa in young children with cystic fibrosis. *The Journal of Infectious Diseases*, 2001 183, 444-452.

[109] Saiman, L; Ishimoto, K; Lory, S; Prince, A. The effect of piliation and exoproduct expression on the adherence of Pseudomonas aeruginosa to respiratory epithelial monolayers. *The Journal of Infectious Diseases*, 1990 161, 541-548.

[110] Saiman, L; Prince, A. Pseudomonas aeruginosa pili bind to asialoGM1 which is increased on the surface of cystic fibrosis epithelial cells. *The Journal of Clinical Investigation*, 1993 92, 1875-1880.

[111] Gupta, SK; Berk, RS; Masinick, S; Hazlett, LD. Pili and lipopolysaccharide of Pseudomonas aeruginosa bind to the glycolipid asialo GM1. *Infection and Immunity*, 1994 62, 4572-4579.

[112] Feldman, M; Bryan, R; Rajan, S; Scheffler, L; Brunnert, S; Tang, H; et al. Role of flagella in pathogenesis of Pseudomonas aeruginosa pulmonary infection. *Infection and Immunity*, 1998 66, 43-51.

[113] Adamo, R; Sokol, S; Soong, G; Gomez, MI; Prince, A. Pseudomonas aeruginosa flagella activate airway epithelial cells through asialoGM1 and toll-like receptor 2 as well as toll-like receptor 5. *American Journal of Respiratory Cell and Molecular Biology*, 2004 30, 627-634.

[114] Miller, SI; Ernst, RK; Bader, MW. LPS, TLR4 and infectious disease diversity. *Nature Reviews Microbiology*, 2005 3, 36-46.

[115] Roger, P; Puchelle, E; Bajolet-Laudinat, O; Tournier, JM; Debordeaux, C; Plotkowski, MC; et al. Fibronectin and alpha5beta1 integrin mediate binding of Pseudomonas aeruginosa to repairing airway epithelium. *European Respiratory Journal*, 1999 13, 1301-1309.

[116] Comolli, JC; Waite, LL; Mostov, KE; Engel, JN. Pili binding to asialo-GM1 on epithelial cells can mediate cytotoxicity or bacterial internalization by Pseudomonas aeruginosa. *Infection and Immunity*, 1999 67, 3207-3214.

[117] Jendrossek, V; Fillon, S; Belka, C; Muller, I; Puttkammer, B; Lang, F. Apoptotic response of Chang cells to infection with Pseudomonas aeruginosa strains PAK and

PAO-I: molecular ordering of the apoptosis signaling cascade and role of type IV pili. *Infection and Immunity*, 2003 71, 2665-2673.

[118] Arora, SK; Ritchings, BW; Almira, EC; Lory, S; Ramphal, R. The Pseudomonas aeruginosa flagellar cap protein, FliD, is responsible for mucin adhesion. *Infection and Immunity*, 1998 66, 1000-1007.

[119] Ernst, RK; Yi, EC; Guo, L; Lim, KB; Burns, JL; Hackett, M; et al. Specific lipopolysaccharide found in cystic fibrosis airway Pseudomonas aeruginosa. *Science*, 1999 286, 1561-1565.

[120] Garrett, ES; Perlegas, D; Wozniak, DJ. Negative control of flagellum synthesis in Pseudomonas aeruginosa is modulated by the alternative sigma factor AlgT (AlgU). *The Journal of Bacteriology*, 1999 181, 7401-7404.

[121] Xia, B; Sachdev, GP; Cummings, RD. Pseudomonas aeruginosa mucoid strain 8830 binds glycans containing the sialyl-Lewis x epitope. *Glycoconjugate Journal*, 2007 24, 87-95.

[122] Flo, TH; Ryan, L; Latz, E; Takeuchi, O; Monks, BG; Lien, E; et al. Involvement of toll-like receptor (TLR) 2 and TLR4 in cell activation by mannuronic acid polymers. *The Journal of Biological Chemistry*, 2002 277, 35489-35495.

[123] Harder, J; Meyer-Hoffert, U; Teran, LM; Schwichtenberg, L; Bartels, J; Maune, S; et al. Mucoid Pseudomonas aeruginosa, TNF-alpha, and IL-1beta, but not IL-6, induce human beta-defensin-2 in respiratory epithelia. *American Journal of Respiratory Cell and Molecular Biology*, 2000 22, 714-721.

[124] Goldman, MJ; Anderson, GM; Stolzenberg, ED; Kari, UP; Zasloff, M; Wilson, JM. Human beta-defensin-1 is a salt-sensitive antibiotic in lung that is inactivated in cystic fibrosis. *Cell*, 1997 88, 553-560.

[125] Taggart, CC; Greene, CM; Smith, SG; Levine, RL; McCray, PB,Jr; O'Neill, S; et al. Inactivation of human beta-defensins 2 and 3 by elastolytic cathepsins. *The Journal of Immunology*, 2003 171, 931-937.

[126] Soong, G; Reddy, B; Sokol, S; Adamo, R; Prince, A. TLR2 is mobilized into an apical lipid raft receptor complex to signal infection in airway epithelial cells. *The Journal of Clinical Investigation*, 2004 113, 1482-1489.

[127] de Bentzmann, S; Roger, P; Dupuit, F; Bajolet-Laudinat, O; Fuchey, C; Plotkowski, MC; et al. Asialo GM1 is a receptor for Pseudomonas aeruginosa adherence to regenerating respiratory epithelial cells. *Infection and Immunity*, 1996 64, 1582-1588.

[128] Mariathasan, S; Monack, DM. Inflammasome adaptors and sensors: intracellular regulators of infection and inflammation. *Nature Reviews Immunology*, 2007 7, 31-40.

[129] Soong, G; Parker, D; Magargee, M; Prince, AS. The type III toxins of Pseudomonas aeruginosa disrupt epithelial barrier function. *The Journal of Bacteriology*, 2008 190, 2814-2821.

[130] Sadikot, RT; Blackwell, TS; Christman, JW; Prince, AS. Pathogen-host interactions in Pseudomonas aeruginosa pneumonia. *American Journal of Respiratory and Critical Care Medicine*, 2005 171, 1209-1223.

[131] Zulianello, L; Canard, C; Kohler, T; Caille, D; Lacroix, JS; Meda, P. Rhamnolipids are virulence factors that promote early infiltration of primary human airway epithelia by Pseudomonas aeruginosa. *Infection and Immunity*, 2006 74, 3134-3147.

[132] Yates, SP; Merrill, AR. A catalytic loop within Pseudomonas aeruginosa exotoxin A modulates its transferase activity. *The Journal of Biological Chemistry*, 2001 276, 35029-35036.

[133] Lau, GW; Hassett, DJ; Ran, H; Kong, F. The role of pyocyanin in Pseudomonas aeruginosa infection. *Trends in Molecular Medicine*, 2004 10, 599-606.

[134] O'Malley, YQ; Reszka, KJ; Rasmussen, GT; Abdalla, MY; Denning, GM; Britigan, BE. The Pseudomonas secretory product pyocyanin inhibits catalase activity in human lung epithelial cells. *American Journal of Physiology – Lung Cellular and Molecular Biology*, 2003 285, L1077-86.

[135] O'Malley, YQ; Reszka, KJ; Spitz, DR; Denning, GM; Britigan, BE. Pseudomonas aeruginosa pyocyanin directly oxidizes glutathione and decreases its levels in airway epithelial cells. *American Journal of Physiology – Lung Cellular and Molecular Biology*, 2004 287, L94-103.

[136] Salvesen, GS; Riedl, SJ. Caspase mechanisms. *Advances in Experimental Medicine and Biology*, 2008 615, 13-23.

[137] Rajan, S; Cacalano, G; Bryan, R; Ratner, AJ; Sontich, CU; van Heerckeren, A; et al. Pseudomonas aeruginosa induction of apoptosis in respiratory epithelial cells: analysis of the effects of cystic fibrosis transmembrane conductance regulator dysfunction and bacterial virulence factors. *American Journal of Respiratory Cell and Molecular Biology*, 2000 23, 304-312.

[138] Plotkowski, MC; Povoa, HC; Zahm, JM; Lizard, G; Pereira, GM; Tournier, JM; et al. Early mitochondrial dysfunction, superoxide anion production, and DNA degradation are associated with non-apoptotic death of human airway epithelial cells induced by Pseudomonas aeruginosa exotoxin A. American *Journal of Respiratory Cell and Molecular Biology*, 2002 26, 617-626.

[139] Jia, J; Wang, Y; Zhou, L; Jin, S. Expression of Pseudomonas aeruginosa toxin ExoS effectively induces apoptosis in host cells. *Infection and Immunity*, 2006 74, 6557-6570.

[140] Shafikhani, SH; Morales, C; Engel, J. The Pseudomonas aeruginosa type III secreted toxin ExoT is necessary and sufficient to induce apoptosis in epithelial cells. *Cellular Microbiology*, 2008 10, 994-1007.

[141] Pederson, KJ; Vallis, AJ; Aktories, K; Frank, DW; Barbieri, JT. The amino-terminal domain of Pseudomonas aeruginosa ExoS disrupts actin filaments via small-molecular-weight GTP-binding proteins. *Molecular Microbiology*, 1999 32, 393-401.

[142] Aktories, K. Bacterial toxins that target Rho proteins. *The Journal of Clinical Investigation*, 1997 99, 827-829.

[143] Kaufman, MR; Jia, J; Zeng, L; Ha, U; Chow, M; Jin, S. Pseudomonas aeruginosa mediated apoptosis requires the ADP-ribosylating activity of exoS. *Microbiology*, 2000 146, 2531-2541.

[144] Jendrossek, V; Grassme, H; Mueller, I; Lang, F; Gulbins, E. Pseudomonas aeruginosa-induced apoptosis involves mitochondria and stress-activated protein kinases. *Infection and Immunity*, 2001 69, 2675-2683.

[145] Jia, J; Alaoui-El-Azher, M; Chow, M; Chambers, TC; Baker, H; Jin, S. c-Jun NH2-terminal kinase-mediated signaling is essential for Pseudomonas aeruginosa ExoS-induced apoptosis. *Infection and Immunity*, 2003 71, 3361-3370.

[146] Kuwano, K. Epithelial cell apoptosis and lung remodeling. *Cellular and Molecular Immunology*, 2007 4, 419-429.

[147] Cannon, CL; Kowalski, MP; Stopak, KS; Pier, GB. Pseudomonas aeruginosa-induced apoptosis is defective in respiratory epithelial cells expressing mutant cystic fibrosis transmembrane conductance regulator. *American Journal of Respiratory Cell and Molecular Biology*, 2003 29, 188-197.

[148] Grassme, H; Kirschnek, S; Riethmueller, J; Riehle, A; von Kurthy, G; Lang, F; et al. CD95/CD95 ligand interactions on epithelial cells in host defense to Pseudomonas aeruginosa. *Science*, 2000 290, 527-530.

[149] Medzhitov, R. Recognition of microorganisms and activation of the immune response. *Nature*, 2007 449, 819-826.

[150] Lee, MS; Kim, YJ. Signaling pathways downstream of pattern-recognition receptors and their cross talk. *Annual Review of Biochemistry*, 2007 76, 447-480.

[151] Campodonico, VL; Gadjeva, M; Paradis-Bleau, C; Uluer, A; Pier, GB. Airway epithelial control of Pseudomonas aeruginosa infection in cystic fibrosis. *Trends in Molecular Medicine*, 2008 14, 120-133.

[152] Skerrett, SJ; Wilson, CB; Liggitt, HD; Hajjar, AM. Redundant Toll-like receptor signaling in the pulmonary host response to Pseudomonas aeruginosa. *American Journal of Physiology - Lung Cellular and Molecular Physiology*, 2007 292, L312-22.

[153] Pier, GB. Pseudomonas aeruginosa lipopolysaccharide: a major virulence factor, initiator of inflammation and target for effective immunity. *International Journal of Medical Microbiology*, 2007 297, 277-295.

[154] Epelman, S; Stack, D; Bell, C; Wong, E; Neely, GG; Krutzik, S; et al. Different domains of Pseudomonas aeruginosa exoenzyme S activate distinct TLRs. *The Journal of Immunology*, 2004 173, 2031-2040.

[155] McNamara, N; Khong, A; McKemy, D; Caterina, M; Boyer, J; Julius, D; et al. ATP transduces signals from ASGM1, a glycolipid that functions as a bacterial receptor. *Proceedings of the National Academy of Sciences*, 2001 98, 9086-9091.

[156] Sutterwala, FS; Mijares, LA; Li, L; Ogura, Y; Kazmierczak, BI; Flavell, RA. Immune recognition of Pseudomonas aeruginosa mediated by the IPAF/NLRC4 inflammasome. *The Journal of Experimental Medicine*, 2007 204, 3235-3245.

[157] Fritz, JH; Ferrero, RL; Philpott, DJ; Girardin, SE. Nod-like proteins in immunity, inflammation and disease. *Nature Immunology*, 2006 7, 1250-1257.

[158] Franchi, L; Stoolman, J; Kanneganti, TD; Verma, A; Ramphal, R; Nunez, G. Critical role for Ipaf in Pseudomonas aeruginosa-induced caspase-1 activation. *European Journal of Immunology*, 2007 37, 3030-3039.

[159] Blohmke, CJ; Victor, RE; Hirschfeld, AF; Elias, IM; Hancock, DG; Lane, CR; et al. Innate immunity mediated by TLR5 as a novel antiinflammatory target for cystic fibrosis lung disease. *The Journal of Immunology*, 2008 180, 7764-7773.

[160] Travassos, LH; Carneiro, LA; Girardin, SE; Boneca, IG; Lemos, R; Bozza, MT; et al. Nod1 participates in the innate immune response to Pseudomonas aeruginosa. *The Journal of Biological Chemistry*, 2005 280, 36714-36718.

[161] Martin, TR; Frevert, CW. Innate immunity in the lungs. *Proceedings of the American Thoracic Society*, 2005 2, 403-411.

[162] Wu, M; McClellan, SA; Barrett, RP; Hazlett, LD. Beta-defensin-2 promotes resistance against infection with P. aeruginosa. *The Journal of Immunology*, 2009 182, 1609-1616

[163] Lopez-Boado, YS; Wilson, CL; Parks, WC. Regulation of matrilysin expression in airway epithelial cells by Pseudomonas aeruginosa flagellin. *The Journal of Biological Chemistry*, 2001 276, 41417-41423.

[164] Sadikot, RT; Zeng, H; Joo, M; Everhart, MB; Sherrill, TP; Li, B; et al. Targeted immunomodulation of the NF-kappaB pathway in airway epithelium impacts host defense against Pseudomonas aeruginosa. *The Journal of Immunology*, 2006 176, 4923-4930.

[165] Gosselin, D; DeSanctis, J; Boule, M; Skamene, E; Matouk, C; Radzioch, D. Role of tumor necrosis factor alpha in innate resistance to mouse pulmonary infection with Pseudomonas aeruginosa. *Infection and Immunity*, 1995 63, 3272-3278.

[166] Lee, JH; Del Sorbo, L; Khine, AA; de Azavedo, J; Low, DE; Bell, D; et al. Modulation of bacterial growth by tumor necrosis factor-alpha in vitro and in vivo. *American Journal of Respiratory and Critical Care Medicine*, 2003 168, 1462-1470.

[167] Reiniger, N; Lee, MM; Coleman, FT; Ray, C; Golan, DE; Pier, GB. Resistance to Pseudomonas aeruginosa chronic lung infection requires cystic fibrosis transmembrane conductance regulator-modulated interleukin-1 (IL-1) release and signaling through the IL-1 receptor. *Infection and Immunity*, 2007 75, 1598-1608.

[168] Power, MR; Peng, Y; Maydanski, E; Marshall, JS; Lin, TJ. The development of early host response to Pseudomonas aeruginosa lung infection is critically dependent on myeloid differentiation factor 88 in mice. *The Journal of Biological Chemistry*, 2004 279, 49315-49322.

[169] Schultz, MJ; Rijneveld, AW; Florquin, S; Edwards, CK; Dinarello, CA; van der Poll, T. Role of interleukin-1 in the pulmonary immune response during Pseudomonas aeruginosa pneumonia. *American Journal of Physiology - Lung Cellular and Molecular Physiology*, 2002 282, L285-90.

[170] Jones, SA. Directing transition from innate to acquired immunity: defining a role for IL-6. *The Journal of Immunology*, 2005 175, 3463-3468.

[171] DiMango, E; Zar, HJ; Bryan, R; Prince, A. Diverse Pseudomonas aeruginosa gene products stimulate respiratory epithelial cells to produce interleukin-8. *The Journal of Clinical Investigation*, 1995 96, 2204-2210.

[172] Strieter, RM. Interleukin-8: a very important chemokine of the human airway epithelium. *American Journal of Physiology - Lung Cellular and Molecular Physiology*, 2002 283, L688-9.

[173] Saba, S; Soong, G; Greenberg, S; Prince, A. Bacterial stimulation of epithelial G-CSF and GM-CSF expression promotes PMN survival in CF airways. *American Journal of Respiratory Cell and Molecular Biology*, 2002 27, 561-567.

[174] Pitrak, DL. Effects of granulocyte colony-stimulating factor and granulocyte-macrophage colony-stimulating factor on the bactericidal functions of neutrophils. *Current Opinion in Hematology*, 1997 4, 183-190.

[175] Shibata, Y; Berclaz, PY; Chroneos, ZC; Yoshida, M; Whitsett, JA; Trapnell, BC. GM-CSF regulates alveolar macrophage differentiation and innate immunity in the lung through PU.1. *Immunity*, 2001 15, 557-567.

[176] Berclaz, PY; Shibata, Y; Whitsett, JA; Trapnell, BC. GM-CSF, via PU.1, regulates alveolar macrophage Fcgamma R-mediated phagocytosis and the IL-18/IFN-gamma -mediated molecular connection between innate and adaptive immunity in the lung. *Blood*, 2002 100, 4193-4200.

[177] Ward, PA. Recruitment of inflammatory cells into lung: roles of cytokines, adhesion molecules, and complement. *The Journal of Laboratory and Clinical Medicine*, 1997 129, 400-404.

[178] Basit, A; Reutershan, J; Morris, MA; Solga, M; Rose, CE,Jr; Ley, K. ICAM-1 and LFA-1 play critical roles in LPS-induced neutrophil recruitment into the alveolar space. *American Journal of Physiology - Lung Cellular and Molecular Physiology*, 2006 291, L200-207.

[179] Downey, DG; Bell, SC; Elborn, JS. Neutrophils in cystic fibrosis. *Thorax*, 2009 64, 81-88.

[180] Kurahashi, K; Kajikawa, O; Sawa, T; Ohara, M; Gropper, MA; Frank, DW; et al. Pathogenesis of septic shock in Pseudomonas aeruginosa pneumonia. *The Journal of Clinical Investigation*, 1999 104, 743-750.

[181] Grassme, H; Jendrossek, V; Riehle, A; von Kurthy, G; Berger, J; Schwarz, H; et al. Host defense against Pseudomonas aeruginosa requires ceramide-rich membrane rafts. *Nature Medicine*, 2003 9, 322-330.

[182] Zaas, DW; Duncan, MJ; Li, G; Wright, JR; Abraham, SN. Pseudomonas invasion of type I pneumocytes is dependent on the expression and phosphorylation of caveolin-2. *The Journal of Biological Chemistry*, 2005 280, 4864-4872.

[183] Galle, M; Schotte, P; Haegman, M; Wullaert, A; Yang, HJ; Jin, S; et al. The Pseudomonas aeruginosa Type III secretion system plays a dual role in the regulation of caspase-1 mediated IL-1beta maturation. *Journal of Cellular and Molecular Medicine*, 2008 12, 1767-1776.

[184] Hauser, AR; Cobb, E; Bodi, M; Mariscal, D; Valles, J; Engel, JN; et al. Type III protein secretion is associated with poor clinical outcomes in patients with ventilator-associated pneumonia caused by Pseudomonas aeruginosa. *Critical Care Medicine*, 2002 30, 521-528.

[185] Smith, EE; Buckley, DG; Wu, Z; Saenphimmachak, C; Hoffman, LR; D'Argenio, DA; et al. Genetic adaptation by Pseudomonas aeruginosa to the airways of cystic fibrosis patients. *Proceedings of the National Academy of Sciences*, 2006 103, 8487-8492.

[186] Jelsbak, L; Johansen, HK; Frost, AL; Thogersen, R; Thomsen, LE; Ciofu, O; et al. Molecular epidemiology and dynamics of Pseudomonas aeruginosa populations in lungs of cystic fibrosis patients. *Infection and Immunity*, 2007 75, 2214-2224.

[187] Mena, A; Smith, EE; Burns, JL; Speert, DP; Moskowitz, SM; Perez, JL; et al. Genetic adaptation of Pseudomonas aeruginosa to the airways of cystic fibrosis patients is catalyzed by hypermutation. *The Journal of Bacteriology*, 2008 190; 7910-7917.

[188] Martin, DW; Schurr, MJ; Mudd, MH; Govan, JR; Holloway, BW; Deretic, V. Mechanism of conversion to mucoidy in Pseudomonas aeruginosa infecting cystic fibrosis patients. *Proceedings of the National Academy of Sciences*, 1993 90, 8377-8381.

[189] Simpson, JA; Smith, SE; Dean, RT. Alginate inhibition of the uptake of Pseudomonas aeruginosa by macrophages. *Journal of General Microbiology*, 1988 134, 29-36.

[190] Simpson, JA; Smith, SE; Dean, RT. Alginate may accumulate in cystic fibrosis lung because the enzymatic and free radical capacities of phagocytic cells are inadequate for its degradation. *Biochemistry and Molecular Biology International*, 1993 30, 1021-1034.

[191] Song, Z; Wu, H; Ciofu, O; Kong, KF; Hoiby, N; Rygaard, J; et al. Pseudomonas aeruginosa alginate is refractory to Th1 immune response and impedes host immune clearance in a mouse model of acute lung infection. *Journal of Medical Microbiology*, 2003 52, 731-740.

[192] Mariencheck, WI; Alcorn, JF; Palmer, SM; Wright, JR. Pseudomonas aeruginosa elastase degrades surfactant proteins A and D. *American Journal of Respiratory Cell and Molecular Biology*, 2003 28, 528-537.

[193] Wu, Y; Xu, Z; Henderson, FC; Ryan, AJ; Yahr, TL; Mallampalli, RK. Chronic Pseudomonas aeruginosa infection reduces surfactant levels by inhibiting its biosynthesis. *Cellular Microbiology*, 2007 9, 1062-1072.

[194] Cobb, LM; Mychaleckyj, JC; Wozniak, DJ; Lopez-Boado, YS. Pseudomonas aeruginosa flagellin and alginate elicit very distinct gene expression patterns in airway epithelial cells: implications for cystic fibrosis disease. *The Journal of Immunology*, 2004 173, 5659-5670.

[195] Williams, P; Camara, M. Quorum sensing and environmental adaptation in Pseudomonas aeruginosa: a tale of regulatory networks and multifunctional signal molecules. *Current Opinion in Microbiology*, 2009 12, 182-191.

[196] Williams, SC; Patterson, EK; Carty, NL; Griswold, JA; Hamood, AN; Rumbaugh, KP. Pseudomonas aeruginosa autoinducer enters and functions in mammalian cells. *The Journal of Bacteriology*, 2004 186, 2281-2287.

[197] Smith, RS; Harris, SG; Phipps, R; Iglewski, B. The Pseudomonas aeruginosa quorum-sensing molecule N-(3-oxododecanoyl)homoserine lactone contributes to virulence and induces inflammation in vivo. *The Journal of Bacteriology*, 2002 184, 1132-1139.

[198] Mahenthiralingam, E; Campbell, ME; Foster, J; Lam, JS; Speert, DP. Random amplified polymorphic DNA typing of Pseudomonas aeruginosa isolates recovered from patients with cystic fibrosis. *Journal of Clinical Microbiology*, 1996 34, 1129-1135.

[199] Wolfgang, MC; Kulasekara, BR; Liang, X; Boyd, D; Wu, K; Yang, Q; et al. Conservation of genome content and virulence determinants among clinical and environmental isolates of Pseudomonas aeruginosa. *Proceedings of the National Academy of Sciences*, 2003 100, 8484-8489.

[200] Hawdon N, Sadeghi Aval P, Barnes R, Gravelle S, Rosengren J, Khan S, Ciofu O, Johansen H, Høiby N, and Ulanova M. 2010. Cellular responses of A549 alveolar epithelial cells to serially collected Pseudomonas aeruginosa from cystic fibrosis patients at different stages of pulmonary infection. *FEMS Immunology and Medical Microbiology*, 2010 59, 207-220.

In: Molecular Aspects of Infectious Diseases
Editor: Mazen T. Saleh
ISBN: 978-1-61728-690-2

Chapter VIII

Prions

Chongsuk Ryou* and Charles E. Mays
Sanders-Brown Center on Aging, Department of Microbiology, Immunology and Molecular Genetics, College of Medicine, University of Kentucky, 800 Rose St. HSRB-326, Lexington, KY 40536, U.S.A.

Abstract

Prions are infectious proteinaceous particles that lack nucleic acid. These agents are responsible for fatal degenerative diseases of the central nervous system in a number of mammalian hosts including humans. Because prions are devoid of nucleic acids unlike other pathogens, they are incapable of reproduction but can self-propagate through the mechanism of protein misfolding. The misfolded protein associated with transmission of disease is a conformationally altered isoform of host-encoded cellular prion protein (PrPC) termed PrPSc. Since PrPSc is resistant to the treatment sufficient to abolish genetic material included in other agents, prions composed of this aberrantly folded isoform of prion protein can not be inactivated by conventional sterilization methods. Studies conducted during the last several decades have provided the information that can resolve the mystery of the nature of prions and unconventional characteristics of prion transmission. In this chapter, we summarize the discovery of prions and the current knowledge of their composition, structure, and replication.

Introduction

Prions are small infectious proteinaceous particles that cause a number of neurodegenerative disorders termed prion diseases or transmissible spongiform encephalopathies (TSEs) [1]. Prion diseases include Creutzfeldt-Jakob disease in human, bovine spongiform encephalopathy in cattle, and scrapie in sheep [1]. Before prions were known, the causative agents for these diseases were thought to be "slow viruses" [2] or

*Corresponding author: Tel (859) 257 4016, Fax (859) 257 8382, E-mail cryou2@email.uky.edu

"unconventional viruses" [3] because prion diseases, in general, are characterized by the lengthy incubation periods and their unusual physico-chemical properties. Search for a virus causing prion disease was not successful. Instead, early scientists learned that the infectious agents retain highly unconventional properties differing from those of viruses. Inactivation studies conducted during mid-1960s to early 1980s demonstrated that infectivity of the agents could be modified by the procedures that destroy proteins, but it was not inactivated by the procedures that destroy nucleic acids. This implicates that the infectious agents for prion diseases are unlikely to be a virus that includes nucleic acids, but more likely to be a novel particle composed exclusively of proteins, for which Stanley Prusiner coined the term 'prion' (proteinaceous infectious particle) [4]. The isolation of prions and identification of these novel pathogens as causative agents for prion diseases were incredibly difficult because the hydrophobic nature of the agent complicated its purification and the bioassays required to study infectivity were time-consuming. Despite these hindrances, much progress has been made over the last several decades.

Infectious but Proteinaceous Pathogens

Early studies characterized the scrapie agent as having several unusual, rather novel, properties distinguishable from other pathogens such as bacteria, fungi, viruses, or viroids. Unlike other infectious agents, prions are extremely resistant to chemical and physical treatments commonly used to inactivate conventional pathogens [1]. Scrapie agents had extraordinary resistance to formalin, a known potent anti-microbial agent [5]. Rigorous treatment at high temperatures such as autoclaving at 121 °C was insufficient to completely destroy prion infectivity [3, 6, 7]. Despite intensive exposure of the infected brain homogenates with UV irradiation, the infectivity of prions was not reduced [4, 8, 9]. Furthermore, in the studies using ionizing radiation, the size of the causative agent was estimated to be $\sim 10^5$ Da which appeared to be too small to carry enough information for its replication as a nucleic acid virus [8, 10-13]. In addition, treatment with nucleases, sonication, and ions that destroy nucleic acids did not alter scrapie infectivity [14-16]. These results suggest that the agent includes no or non-functional nucleic acids, which led to the interesting proposition that scrapie was caused by the self-replication of a protein independently of nucleic acid [4, 17].

Meanwhile, the concept of self-replicating proteinaceous agents was continuously challenged and the studies to identify a virus or viroid responsible for scrapie were ongoing. To directly discern whether or not the scrapie agent contained a virus, β-propiolactone, an alkylating compound known to inactivate a number of viruses, was mixed and incubated with brain homogenate of mice infected by mouse-adapted scrapie [18]. However, this failed to prevent transmission. Phenol extraction methods used to isolate RNA and DNA of viruses and viroids failed to extract scrapie-associated nucleic acid [19, 20]. Additionally, several characteristics of prion diseases were distinctly different from those expected for a nucleic acid-associated disease such as the lack of an inflammatory response in the brain, no change in the number of white blood cells in the cerebrospinal fluid, and no recovery or remission of infection [3]. Collectively, these findings suggest that the infectious agent causing prion diseases is unlikely to be a virus or viroid.

On the other hand, some procedures using strong detergents and protein denaturants such as SDS, NaOH, chaotropic guanidinium salts, phenol, urea, or free chlorine, effectively reduced prion infectivity [21]. Furthermore, treatments with 1-2% chlorine, extended autoclaving above 132 °C, or NaOH denaturation with additional autoclaving were efficient to inactivate prions and useful for decontamination [22]. Additionally, prion infectivity was dose-dependently inactivated by diethyl pyrocarbonate, a compound that rapidly carbethoxylates proteins but not nucleic acid residues. Then, hydroxylamine was shown to reverse the effect of diethyl pyrocarbonate by protein-specific decarbethoxylation [16, 23-25]. Together these studies indicate that the proteinaceous nature of scrapie agents (prions) confers their infectivity.

Isolation of Infectious Agents Composed of Proteins only

In the search for proteinaceous agents causing prion disease, a hydrophobic protein, later termed scrapie-associated prion protein (PrPSc), was isolated from the brains of prion-infected animals [26] using a combination of detergent extraction, limited digestions with proteases and nucleases, differential centrifugation, Sarkosyl agarose gel electrophoresis, and sucrose gradient centrifugation [25, 27-32]. These procedures enriched a single protein, which is insoluble under mild detergent conditions, by >100-fold and eliminated >98% of the other proteins and polynucleotides present in the sample. Analysis of the highly purified samples with sensitive methods showed that nucleic acid contaminants were composed of small pieces of nucleic acids less than 80 nucleotides long [33] and estimated the nucleic acid concentration to be less than one molecule per infectious unit, a minimal prion dose causing disease in an animal [34].

Analysis of the concentrated prion preparation with SDS polyacrylamide gel electrophoresis identified a protease-resistant protein with a molecular size of 27-30 kDa, termed PrP27-30 [35, 36]. Highly purified PrP27-30 retained an infectivity titer proportional to its concentration. However, infectivity of PrP27-30 could be decreased by denaturation with strong chaotropic guanidium salts and anionic detergent SDS [14]. Similar to the ineffectiveness of UV-irradiation on the crude scrapie preparations, the purified PrP27-30 irradiated by UV also resisted to reduce infectivity [13]. Production of a rabbit anti-serum against PrP27-30 [37] revealed that PrP27-30 was derived from a precursor protein with the molecular weight of 33-35 kDa (PrP33-35), which was produced in both normal and scrapie-infected brains. Scrapie-associated PrP33-35 resulted in PrP27-30 after proteolytic digestion, whereas protease-sensitive, cellular PrP33-35 found in normal tissue failed to produce a resistant product [38].

Purification of PrPSc, confirmation of its infectivity, assessment of the role of nucleic acids in its infectivity, and generation of antibody that recognize this protein provided data to support that prions propagate without the replication of nucleic acids, and prion infectivity does not depend on a viral or viroid genome. Moreover, these data demonstrated that the major protein composed of the infectious prion particle is PrP27-30, which is partially resistant to protease digestion and post-translationally modified from the host PrP33-35 [1, 39].

Identification of Prp Gene

Determination of the N-terminal sequence of PrP27-30 [40] followed by molecular cloning of cDNA identified a single gene encoding PrP27-30 in hamsters, mice, and humans [38, 41]. The prion protein (PrP) gene is highly conserved in many mammalian species [42]. DNA sequence determination of the identified cDNA clones revealed that PrP27-30 represents a fragment of a larger cellular protein termed PrP [43]. The entire open reading frame of PrP was found to be uninterrupted within a single exon, indicating that PrP27-30 is not created by alternative RNA splicing [43, 44]. Supporting that PrP27-30 is a product of a host-encoded protein, PrP-related mRNA sequences were found in the brains of both normal and scrapie-infected animals, as well as in many other tissues [38, 41, 43]. The PrP gene is highly and widely expressed in the embryo during prenatal periods [38, 41, 45]. In the adult, expression of the PrP gene is still ubiquitous in almost all tissues but most abundant in the brain [46]. Within the brain, expression of the PrP gene is at the highest concentration in neurons, but not limited to them [38, 41, 45]. These data provided the direct evidence that PrP27-30, the major protein in the prion particle, originates from a host-encoded protein.

Protein-only Hypothesis

As the non-viral nature of scrapie agents was suggested by the UV irradiation studies [8], Griffith proposed the possibility that the causative agent for disease might be a protein that is able to replicate in a nucleic acid free manner [17]. This launched a series of seminal works, including Prusiner's, in the following years, which led to the introduction of the protein-only hypothesis [4]. The protein-only hypothesis postulates that a protein (i.e. prion) with a certain conformation can replicate without transfer of genetic information from DNA or RNA. This contradicts the 'central dogma of molecular biology', which describes that genetic information flows from DNA to RNA to proteins. However, as shown in non-Mendelian type inheritance of phenotypic traits in yeasts [47], protein-only hypothesis has become widely accepted as a novel paradigm for conveying molecular information and explains many phenomena involved in prion diseases of mammals. According to the protein-only hypothesis, prions, composed exclusively of PrP^{Sc}, are responsible for infectivity and are replicated through conversion of cellular prion protein (PrP^{C}) to PrP^{Sc} in a nucleic acid free manner during transmission of prion diseases. As an alternative, a viral hypothesis for prion diseases has been proposed. According to this hypothesis, prion diseases are caused by an unidentified "slow" virus or a replicable informational molecule such as a nucleic acid bound to PrP. This hypothesis is based on the observation that virus-like particles have been found in some of the Creutzfeldt-Jakob disease (CJD) or scrapie prion-infected cells in culture [48, 49]. Evidence in favor of each hypothesis is summarized in Table 1.

Recent observations have provided strong evidence in favor of the protein-only hypothesis. Synthetic amyloid fibrils spontaneously generated from truncated recombinant PrP in the test tubes demonstrated an "infectious" nature by causing disease in transgenic mice expressing truncated PrP^{C} [50]. Furthermore, cell-free PrP^{C} conversion conducted with PrP^{Sc} purified from prion-infected brains and radio-labeled PrP^{C} derived from cultured cells demonstrated formation of PrP^{Sc}-like molecules from the radio-labeled PrP^{C} [51]. More recently, protein misfolding cyclic amplification using brain homogenate of healthy and prion-infected animals showed generation of a large amount of PrPSc from a minute amount of PrPSc [52]. The amplified PrPSc preserved infectivity and other features of the original PrPSc including clinical and biochemical properties [53]. Although none of the in vitro prion propagation assays available perfectly simulate the event occurring in nature [54], these data seem to reflect a certain aspect of prion replication proposed by the protein-only hypothesis.

Table 8.1. Evidence in favor of protein-only and viral hypotheses.

Protein-only hypothesis	Viral hypothesis
Any known pathogens such as viruses, bacteria, or fungi have not been isolated to confirm conclusive associated with prion diseases.	Biological properties involved in the strain variation shown in many prion diseases resemble the strain variation seen by RNA viruses.
Association of nucleic acid with infectivity has not been definitely established.	Prion strain variances can not be explained whether such variance in PrP^{Sc} conformation are the cause, or simply the result, of the prion strain phenomenon.
The causative agent is resistant to treatment with nucleases [14-16].	The long incubation and rapid onset of symptoms resemble some lentiviral infections.
The concept for existence of genetic material specific for prion pathogen (PrP^{Sc}) is unexplainable because amino acid sequence of PrP^{Sc} and PrP^{C} are identical.	Infectivity does not correlate with the presence or absence of PrP^{Sc} (Some prions are infectious with no detectable PrP^{Sc}, but other prions are not infectious even in the presence of abundant PrP^{Sc} [134, 135]).
The mutations in the PrP gene cause genetic prion diseases with no prion infection [133].	A small quantity of nucleic acids is present in highly purified PrP^{Sc}.
PrP^{C} –deficient animals do not develop prion diseases [82-84].	PrP^{Sc} interacts with RNA that enhances to catalyze the conversion of PrP^{C} into PrP^{Sc} in vitro [136, 137]
The donor PrP^{Sc} is replicated by PrP^{C} of the recipients: The propagated PrP^{Sc} retains the amino acid sequence of the recipient species during transmission between species.	
Prion infection does not induce immune response in the host, unlike infections by other pathogens.	

Isoforms of Prion Protein: PrP^{C} and PrP^{sc}

The PrP gene encoding PrP^{C} is determined to locate on the gene locus *PRNP* in the genome of hosts [1]. PrP gene orthologs have been identified in a number of animal species. For example, *PRNP* is assigned to human chromosome 20 and the corresponding mouse chromosome 2 [55]. Human PrP gene includes a single open reading frame encoding a polypeptide composed of 253 amino acid residues (Figure 8.1) [1]. PrP has a hydrophobic, amino-terminal leader sequence that directs translocation of the polypeptide into the lumen of the endoplasmic reticulum and eventually presents the polypeptide on the cell surface. PrP also carries a signal sequence for the attachment of glycosylphosphatidylinositol (GPI) anchor at the carboxyl-terminus, which directs to post-translationally tether the polypeptide to the plasma membrane [56].

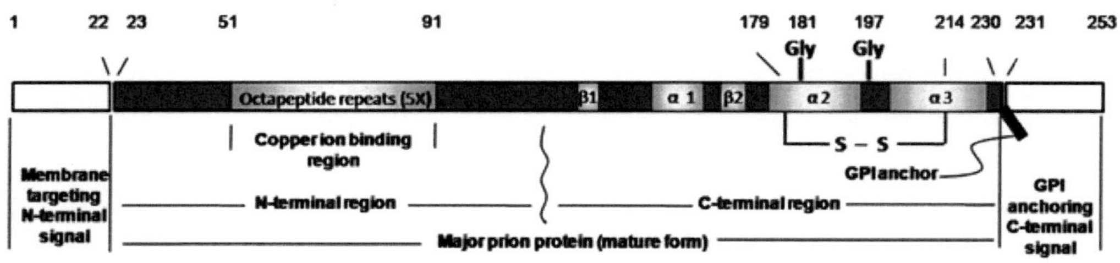

A

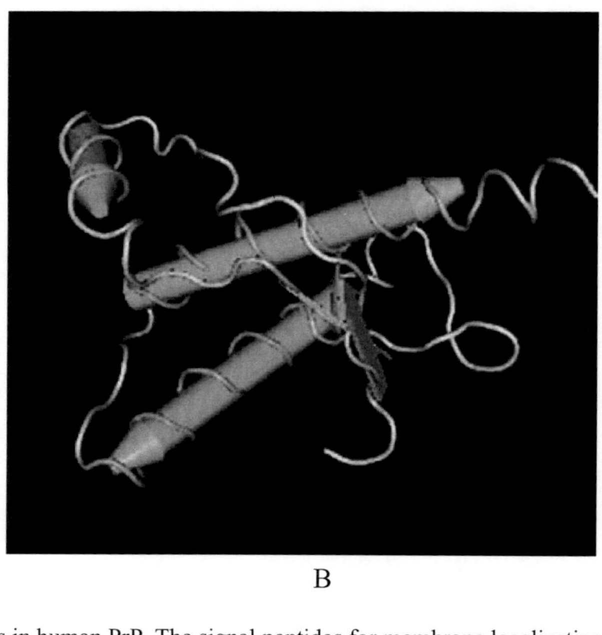

B

Figure 8.1. (A) Domains in human PrP. The signal peptides for membrane localization and glycosylphosphatidylinositol (GPI) anchoring, major prion protein after processing and maturation, octapeptide repeats, three α-helices (α1, α2, and α3), two β-sheets (β1 and β2), glycosylation (Gly) sites, disulfide bridges were shown with the amino acid numbering based on the registered human PrP sequence (UniProtKB/Swiss-Prot entry P04156). (B) NMR structure of C-terminal region of human PrP at pH 7.0 [132]. The image (MMDB ID:24910, PDB ID: 1HJM) obtained from the NCBI-Structure [138]. Rods; α helices. Arrows: β sheets.

PrP includes two sites for N-linked glycosylation at amino acid residues 181 and 197, and two cysteines for a disulfide bridge at amino acid residues 179 and 214. Finally, PrP contains the octarepeat region composed of five repeats of highly conserved eight amino acid sequences in its amino-terminus, which is known to interact with copper ions for a yet to be discovered function [57]. After processing, mature PrP^C is found at the cell surface as a GPI anchored protein of 209 amino acids with a disulfide bond and various glycosylation states. PrP^C can be released from the plasma membrane when treated with phosphatidylinositol-specific phospholipase C that cleaves the GPI anchor from the carboxyl-terminus of the polypeptide [56, 58]. PrP^C is present as a mixture of di-, mono- or un-glycosylated protein [59]. During biogenesis in the cells, PrP^C synthesized in the endoplasmic reticulum is trafficked through Golgi and eventually displayed on the plasma membrane [60]. In cultured neuronal cells, PrP^C is concentrated in lipid-rich domains of plasma membrane known as cavelolae-like domains [1, 60].

PrP^C is the precursor molecule for infectious, disease-associated PrP^{Sc}. During prion infection, the full-length PrP^C is aberrantly folded into PrP^{Sc} by a poorly understood conformational misfolding process. Treatment with protease removes nearly 67 amino acids in the amino-terminus of PrP^{Sc} and generates PrP27-30, the protease-resistant core of PrP^{Sc} that was first isolated as described above [35, 36]. Both isoforms of PrP are identical in their amino acid sequences and state of modifications. However, they differ in their three dimensional conformations and exhibit completely different biochemical and biophysical properties (Table 2). PrP^C is protease-sensitive and soluble in detergents, whereas PrP^{Sc} is hydrophobic, and insoluble in the non-denaturing detergents. Also, PrP^{Sc} is lipophilic, therefore, non-covalently associated with lipids such as sphingomyelin and cholesterol [61].

Table 8.2. Different properties of PrP^C and PrP^{Sc}.

PrP^C	PrP^{Sc}
Cellular protein	Disease-associated protein
Not infectious	Infectious; pathogenic
Normal folding	Misfolded
High α-helical contents	Increased β-sheet contents
Easily soluble (hydrophilic)	Insoluble (hydrophobic, lipophilic)
Sensitive to protease treatment	Partially, but highly resistant to protease

PrP^{Sc} tends to aggregate to form fibrillar structures that favorably bind to amyloidophilic compounds such as Congo Red and thioflavin T [62]. Although not found in every prion-infected specimen, aggregated structures of PrP^{Sc}, known as scrapie associated fibrils (SAF), have been observed using electron microscopy by examining subfractions of brain from prion-affected animals [63]. SAF has been consistently seen not only in multiple prion strains naturally found in sheep, cervids, and human patients, but also in a variety of strains experimentally transmitted to laboratory animals [63-66]. Morphological properties of all SAF fit those of the β-pleated protein fibrils associated with amyloidosis. However, SAF are distinguishable from amyloid fibrils and paired helical filaments associated with brain samples from other amyloidogenic disorders such as Alzheimer's disease, Parkinson's disease, and amyotrophic lateral sclerosis [64]. Bioassay studies demonstrated a correlation between SAF and prion infectivity: the prion titer of SAF diminished when purified SAF fractions were boiled in the presence of 1-2% SDS [64, 67]. Antibodies to PrP27-30 were shown to recognize the SAF, which confirmed that these fibrils were packed with thousands of PrP molecules [65].

PrP^c Function in Physiology and Prion Pathogenesis

The role of PrP^C in physiology remains ambiguous because a number of different lines of PrP knockout mice have not shown any major impairments or distinct alterations in physiology and behavior [45, 68-71]. However, expression of PrP^C in a tissue- and time-specific manner, as well as localization of PrP^C on the lipid rafts, implicates a certain biological function of PrP^C. Although the function of PrP^C has not been confirmed, it appears to play a role in lymphocyte activation [72], synaptic plasticity [73], neuroprotection [71, 74], signal transduction [75], and metabolic functions related to copper binding properties [76]. Furthermore, recent studies showed that PrP^C is involved in long-term renewal in hematopoietic stem cells [77] and differentiation of neural stem cells for neurogenesis [78].

The role of PrP^C in the pathogenesis of prion disease is well-defined. Development of several genetically manipulated murine models has demonstrated that PrP gene dosage influences the rate of disease progression. In transgenic mouse lines such as Tg(MoPrP-A)4053/FVB, Tga20, and Tg(MoPrP)4112, six- to ten-fold excess expression of PrP^C drastically reduced the incubation period [79-81]. Moreover, animals null for the *PRNP* gene are completely resistant to disease with no clinical and pathological signs [82-84]. These studies suggest that incubation period and PrP^C titer are inversely associated. Additionally, distinct polymorphisms within the PrP^C sequence are critical in prion pathogenesis. For instance, polymorphisms at residues 108 and 189 in a variety of mouse strains demonstrated the ability of the gene, denoted $Prnp^a$ for the short incubation gene and $Prnp^b$ for the long incubation gene, to dictate the incubation period [44, 85-88]. These findings correlated with earlier work by Dickinson and colleagues that predicted variations in incubation periods of different mice strains with scrapie were due to the *PRNP* gene [89-91]. In addition, most sheep predisposed to natural scrapie have a polymorphism at codons 136, 154, or 171 [92, 93]. Furthermore, all heritable TSEs of humans have been genetically linked to specific mutations in PrP, and the clinical presentation of BSE in humans as variant CJD has only been recognized in individuals homozygous for methionine at codon 129 [94, 95].

Structure of PrPc and PrP^{sc}

The primary structures of PrP^C and PrP^{Sc} are identical, but the two isoforms differ in many biochemical and biophysical properties. Such differences are believed to be a consequence of a different secondary and tertiary structure of both PrP^C and PrP^{Sc}. The secondary structure of both PrP isoforms has been analyzed by Fourier transform infrared (FTIR) and circular dichroism (CD) spectroscopy in aqueous media. FITR and CD spectroscopy studies demonstrated that PrP^C has high α-helix (42%) and relatively little β-sheet (3%) contents [96]. However, the same analysis showed that full-length PrP^{Sc} had dramatically increased β-sheet (~40%) and moderately decreased α-helix (25%) contents [96, 97], while the proteinase K resistant core, PrP27-30, had elevated β-sheet content (~50%) with a lower percentage of α-helices (< 15%) [96-98]. The increase of β-sheet contents in

PrP^{Sc} suggest that the fundamental event in generating the disease-associated, infectious PrP isoform is conformational conversion of α-helices into β-sheets.

The high resolution three dimensional structure of PrP^C was determined by nuclear magnetic resonance (NMR) spectroscopy. Three dimensional models were typically created for two C-terminal globular domain fragments, PrP(90-230) and PrP(121-230) [99-101]. In general, PrP^C includes three α-helices and two very short antiparallel β-sheets within the carboxyl-terminal globular core [99, 101]. Analysis of full-length PrP(23-230) proposed a long, flexibly disordered tail attached to the N-terminus of the globular domain [100]. The octarepeats included in the unstructured amino-terminus of PrP^C are free in solution and known to bind 4–6 copper ions [102]. This interaction could cooperatively induce the proper structure of PrP^C. Furthermore, PrP^C contains a scaffold created by a disulfide bond bridging the second and third α-helices (α2 and α3). NMR structure based on the sequences of a number of mammalian species revealed two highly substituted significant regions that represent potential species-dependent surface recognition sites for protein-protein interactions. Variable residues within 96-167 on the first α-helix were predicted to provide an adequate site for interaction with PrP^{Sc}. A second binding site was proposed to exist for protein X, a hypothetical cofactor assisting conversion of PrP^C to PrP^{Sc}, at the loop between the second β-strand (β2) and the α2 helix ranging from positions 164-174 along with C-terminal residues 215-223 that are in close proximity [103].

Three dimensional NMR models of PrP^C, together with other lines of evidence such as transgenic animals expressing mutated PrP^C in a specific region, provided insight for the species barrier and the molecular mechanism of conformational conversion of PrP isoforms. In comparing intra-species PrP structures obtained by NMR, plasticity of the protein X binding site within the β2-α2 loop and the α3 helix appears to determine the susceptibility of a given species to a respective TSE [104]. This critical β2-α2 loop of PrP^C is highly structured for bank voles, elk, tammar wallabies, and also quite well-ordered in Syrian hamsters; however, this region is disordered for most other mammalian species including bovine, human, and mouse [104-107]. The structured loop in bank voles and elk is primarily linked to an Asn residue at position 170, which seems to be correlated with either a higher susceptibility to horizontal TSE transmission or a lower interspecies transmission barrier [105, 107]. In contrast, no TSE has been reported thus far in marsupials such as the tammar wallaby. The ability of the tammar wallaby to evade infection with TSEs may be because the β2-α2 loop is stabilized by long-range interactions between residues 166 and 225; not local effects from amino acid substitutions as observed in bank voles and elk [104]. At the binding site for protein X, α3 is extended by approximately seven residues for bovine, human, and Syrian hamster, which could increase susceptibility to TSEs and compensate for the lack of a completely ordered loop [108-110]. Additionally, the protection from TSEs exhibited by canines correlates with the presence of Asp-159 and Arg-177 that causes unique charge distribution patterns in the area surrounding the β2-α2 loop [111]. Although TSEs have only been identified in mammals, PrP^C of the chicken, turtle, and frog are protected from disease by only slight differences in the length and positions of the α-helices and β-strands [112]. The current library of three dimensional models for PrP^C indicates that only minor changes in a specific region of PrP can completely alter the degree of susceptibility to TSEs.

Despite extensive NMR studies, X-ray crystallographic approaches to define the solution structure of PrP have been limited by difficulties in crystallizing the protein. Recently, Haire et al. demonstrated that the globular domain of ovine PrP(123-230) had an overall organization and positioning of the three α-helices and short anti-parallel β-sheet congruent with the other mammalian proteins determined by NMR [113]. In a separate study, structural analysis of PrP was made possible by co-crystallization with anti-PrP antibodies or their Fab fragments [114, 115]. Using the co-crystallized protein, it was determined that the major ovine residues involved in scrapie-sensitivity polymorphisms (136, 154, and 171) are on the protein surface with their side chains exposed, which can destabilize the protein in an additive fashion making them more susceptible to aggregate or change into a pathological conformation [115]. Also, X-ray crystallography of PrP provided fascinating information on dimerization of PrP. Earlier studies suggested that PrP in an α-helical structure with the intact intramolecular disulfide bridge can form dimers [116]. In contrast, the crystal structure of a dimer of monomeric human PrP (120–230) showed that dimerization was formed by the three dimensional swapping of the residues located between α2 and α3, and rearrangement of the disulfide bond to bridge two monomers [117]. As a result, a new antiparallel β-sheet is formed at the dimer interface comprising strands from each of the monomers. If dimerization occurs in vivo, this suggests a potential mechanism for PrP oligomerization, where PrP^{C} dimers could accelerate the formation of PrP^{Sc} [117].

Models of PrP^{Sc} Aggregates

Structure of PrP^{Sc} is not accessible by NMR or X-ray analysis due to its insolubility. Additionally, heavy glycosylation and GPI anchors obscure the successful visual presentation of the protein core from the PrP^{Sc} aggregates. Currently, the structure of PrP^{Sc} is estimated only by simulation. Based on the determination of the secondary structures of PrP^{C} and PrP^{Sc} by CD and FTIR spectroscopy, the initial model of PrP^{Sc} describes a structure stabilized by intermolecular interactions, in which amino-terminus of the PrP core region is changed into β-sheets but α2 and α3 are assumed unchanged [96]. A new approach, negative stain electron microscopy, allowed visualization of two dimensional crystals with an apparent hexagonal lattice that is exclusively in purified fractions containing high titers of prion infectivity [118]. Experimental and computational results define these two dimensional crystals as trimeric structures containing parallel left-handed β-helices [119]. In this β-helix model, the β-helical amino-terminus (residues 89–175) is located inside the hexagonal unit, but α-helical carboxyl-terminus including α2 and α3 is located at the outer surface with the retained glycosyl groups and disulfide linkage. During elongation of a fibril, two trimers could assemble by hydrogen bonds that link the top of one β-helix to the bottom of the next one, which would allow the preserved carboxyl-terminal α-helices of each monomer to stack with their glycosyl moieties facing out [119]. As an alternative, the spiral model has been proposed on the basis of molecular dynamics simulation using PrP with D147N mutation [120, 121]. In this model, all three native α helices remain unaffected but four β-sheets are formed within the residues 116–164 of a PrP monomer that can be assembled into continuous twisting filaments through intermolecular β-sheets. The third model describes the parallel in-register β structure [122, 123]. Hydrogen/deuterium exchange analysis of amyloids generated from recombinant PrP (90-230) indicated the presence of a largely unstructured region at residues 90-168 and a systematically hydrogen-bonded cross-β structure in the carboxyl-terminal regions at 169-213 and 218-224 [122]. Also, site-directed spin labeling and *electron paramagnetic resonance* spectroscopy of synthetic amyloids identified the β-sheet core at residues ~160/170-220 [123]. These studies suggest that parallel, in-register β-sheet structures could be formed at the carboxyl-region as a result of the refolding of original α helices of PrP^C. Based on this observation, it is possible to propose that PrP amyloids appear to be stabilized by interlocking aligned amino acid side chains in the short stretches of stacked β sheet monomers [124]. These models describe PrP^{Sc} conformation formed under specific in vitro conditions, but verification of the correct model will be difficult without development of a method to obtain a high resolution solution structure.

Prion Replication

Conformational alteration of PrP^C to PrP^{Sc} occurs during prion pathogenesis. However, the absence of detailed structural information for PrP^{Sc} has hindered understanding the precise molecular mechanism in which PrP^C is converted into PrP^{Sc}. Two alternative models for prion replication are proposed: the nucleation-dependent model [125] and the template-assisted conversion model [126]. The nucleation-dependent model postulates that PrP^{Sc} seeds form an ordered nucleus, which is a thermodynamically unfavorable event and requires a long lag time before the aggregates are detectable. These ordered multimeric PrP^{Sc} serve as a polymerization nucleus. Once a stable nucleus is formed, it immediately stabilizes and incorporates PrP^{Sc} monomers and/or preformed small PrP^{Sc} multimers within the PrP^{Sc} amyloid. This event is thermodynamically favorable compared to the initial step to form a PrP^{Sc} nucleus. As PrP^{Sc} is propagated, amplification arises from newly formed faces on the outside of the PrP^{Sc} amyloid. Seeding of the PrP^{Sc} nucleus is highly specific and requires complementarities between the seed and the amyloid. The template-assisted conversion model proposes that the misfolded monomeric PrP^{Sc} serves as a template to convert PrP^C to PrP^{Sc} that forms amyloid afterwards. Under normal conditions, PrP^C exists in equilibrium as the dominant state with a conformational intermediate, PrP*. When PrP^{Sc} is available, a monomer of PrP^{Sc} can create a heteromultimer by binding with PrP* that can be spontaneously converted into a homomultimer of PrP^{Sc}. Additionally, an unidentified component, designated protein X, may play a critical role in conversion by acting as an auxiliary factor [127, 128]. In this model, protein X preferentially interacts with PrP^C to induce PrP* by destabilizing PrP^C, or to increase the stability of PrP* for accelerated conversion into PrP^{Sc}. Once the PrP^{Sc} homodimer dissociates, two PrP^{Sc} templates are available to exponentially induce conversion and protein X is free to interact with PrP^C [129]. Thus, repeated interaction of PrP^{Sc} with PrP^C molecules results in PrP^{Sc} replication. Both models require physical association between PrP^C and PrP^{Sc} for the PrP^{Sc} propagation process, while they differ in the role of PrP^{Sc} aggregates during PrP^{Sc} propagation. PrP^{Sc} aggregates are absolutely required for PrP^{Sc} propagation and formation of a PrP^{Sc} nucleus is the rate-limiting step in the nucleated polymerization model. In contrast, PrP^{Sc} aggregates are not essential for PrP conversion in the template-assisted conversion model, but the rate-limiting step is conformational alteration of PrP^C to PrP^{Sc}.

Conclusion

The causative agent of prion diseases has eluded identification for decades because of its unconventional characteristics. Prions are infectious particles primarily composed of a misfolded protein, PrP^{Sc}, which has the ability to self-replicate in the absence of nucleic acids. The proposed replication process is based on the conformational alteration of host-encoded PrP^{C} without undergoing covalent modifications. In fact, recent development from multiple in vitro prion propagation studies has demonstrated that prion replication is facilitated by conversion of PrP^{C} to PrP^{Sc}. However, to prove protein-only hypothesis, a product retaining infectivity congruent to that observed in nature must be generated using synthetic substances in the complete absence of additional material obtained from tissues or cultured cells.

Information on the detailed mechanism of prion conversion is not available. Due to the significant differences at the secondary structures determined by low resolution conformational analysis of PrP^{C} and PrP^{Sc}, it is believed that α-helices of PrP^{C} must be converted into β-sheets to produce the disease-associated isoform. The specific regions undergoing conformational conversion have been predicted by modeling of high resolution three dimensional PrP structures and computational simulation of PrP aggregates. However, reconciliation of current models seems to require additional studies that can provide higher resolution structures and clarification on various conformations of PrP fibrils prepared differently in vitro. Although amyloid fibrils generated in vitro are the most prominent material for the understanding of PrP^{Sc} structure, recent studies indicate that PrP fibril formation is dependent on the conditions used for their generation [130], which raises a question of whether the structure of these prion amyloid fibrils formed in vitro represents the conformation found in nature. Nevertheless, verification of the predicted β-structure region of PrP^{Sc} in these alternate amyloid fibrils should be analyzed to understand the diverse nature of prion fibril formation and the conditions affecting it. Recently, a method to promote the formation of large and well-organized two dimensional prion crystals has been reported [131]. Manipulation of these two dimensional crystals may introduce conditions that are suitable for determining the high-resolution three dimensional structure for PrP^{Sc}. Such effort will aid in elucidating the precise molecular mechanism of the conformational alteration of PrP^{C} to PrP^{Sc}, and eventually provide crucial information on prion propagation and transmission of infectivity.

Acknowledgments

We are grateful to Paula Thomason and Sally Malley for editorial assistance. This work was supported in part by funds from the Sanders-Brown Center on Aging and College of Medicine, University of Kentucky

References

[1] Prusiner SB. Prions. *Proc Natl Acad Sci USA*, 1998, 95, 13363-13383.

[2] Sigurdsson B. Rida, a chronic encephalitis of sheep with general remarks on infections which develop slowly and some of their special characteristics. *Br Vet J*, 1954, 110, 341-354.

[3] Gajdusek DC. Unconventional viruses and the origin and disappearance of kuru. *Science*, 1977, 197, 943-960.

[4] Prusiner SB. Novel proteinaceous infectious particles cause scrapie. *Science*, 1982, 216, 136-144.

[5] Gordon WS. Advances in veterinary research. *Vet Res*, 1946, 58, 516-520.

[6] Stamp JT. Scrapie: a transmissible disease of sheep. *Vet Rec*, 1962, 74, 357-362.

[7] Hunter GD, Millson GC. Studies on the heat stability and chromatographic behavior of the scrapie agent. *J Gen Microbiol*, 1964, 37, 251-258.

[8] Alper T; Cramp WA; Haig DA, Clarke MC. Does the agent of scrapie replicate without nucleic acid? *Nature*, 1967, 214, 764-766.

[9] Gibbs CJ, Jr.; Gajdusek DC, Latarjet R. Unusual resistance to ionizing radiation of the viruses of kuru, Creutzfeldt-Jakob disease. *Proc Natl Acad Sci USA*, 1978, 75, 6268-6270.

[10] Alper T. Scrapie agent unlike viruses in size and susceptibility to inactivation by ionizing or ultraviolet radiation. *Nature*, 1985, 317, 750.

[11] Alper T; Haig DA, Clarke MC. The scrapie agent: evidence against its dependence for replication on intrinsic nucleic acid. *J Gen Virol*, 1978, 41, 503-516.

[12] Alper T; Haig DA, Clarke MC. The exceptionally small size of the scrapie agent. *Biochem Biophys Res Commun*, 1966, 22, 278-284.

[13] Bellinger-Kawahara C; Cleaver JE; Diener TO, Prusiner SB. Purified scrapie prions resist inactivation by UV irradiation. *J Virol*, 1987, 61, 159-166.

[14] Bellinger-Kawahara C; Diener TO; McKinley MP; Groth DF; Smith DR, Prusiner SB. Purified scrapie prions resist inactivation by procedures that hydrolyze, modify, or shear nucleic acids. *Virology*, 1987, 160, 271-274.

[15] McKinley MP; Masiarz FR; Isaacs ST; Hearst JE, Prusiner SB. Resistance of the scrapie agent to inactivation by psoralens. *Photochem Photobiol*, 1983, 37, 539-545.

[16] Prusiner SB; Groth DF; McKinley MP; Cochran SP; Bowman KA, Kasper KC. Thiocyanate and hydroxyl ions inactivate the scrapie agent. *Proc Natl Acad Sci USA*, 1981, 78, 4606-4610.

[17] Griffith JS. Self-replication and scrapie. *Nature*, 1967, 215, 1043-1044.

[18] Haig DA, Clarke MC. The effect of beta-propiolactone on the scrapie agent. *J Gen Virol*, 1968, 3, 281-283.

[19] Marsh RF; Semancik JS; Medappa KC; Hanson RP, Rueckert RR. Scrapie and transmissible mink encephalopathy: search for infectious nucleic acid. *J Virol*, 1974, 13, 993-996.

[20] Ward RL; Porter DD, Stevens JG. Nature of the scrapie agent: evidence against a viroid. *J Virol*, 1974, 14, 1099-1103.

[21] Prusiner SB; Cochran SP; Baringer JR; Groth D; Masiarz F; McKinley M; Bildstein C; Hadlow WJ; Race RE, Eklund CM. *Slow viruses: molecular properties of the agents*

causing scrapie in mice and hamsters. In: Battistin L, Hashim G, Lajtha A, editors. Neurochemistry and Clinical Neurology. New York: Allan R. Liss Publishers; 1980. p. 73-90.
[22] Taylor DM. Inactivation of transmissible degenerative encephalopathy agents: A review. *Vet J*, 2000, 159, 10-17.
[23] Diener TO; McKinley MP, Prusiner SB. Viroids and prions. *Proc Natl Acad Sci USA*, 1982, 79, 5220-5224.
[24] McKinley MP; Masiarz FR, Prusiner SB. Reversible chemical modification of the scrapie agent. *Science*, 1981, 214, 1259-1261.
[25] Prusiner SB; Groth DF; Cochran SP; Masiarz FR; McKinley MP, Martinez HM. Molecular properties, partial purification, and assay by incubation period measurements of the hamster scrapie agent. *Biochemistry*, 1980, 21, 4883-4891.
[26] Prusiner SB; McKinley MP; Groth DF; Bowman KA; Mock NI; Cochran SP, Masiarz FR. Scrapie agent contains a hydrophobic protein. *Proc Natl Acad Sci USA*, 1981, 78, 6675-6679.
[27] Prusiner SB; Garfin DE; Cochran SP; McKinley MP; Groth DF; Hadlow WJ; Race RE, Eklund CM. Experimental scrapie in the mouse: electrophoretic and sedimentation properties of the partially purified agent. *J Neurochem*, 1980, 35, 574-582.
[28] Prusiner SB; Groth DF; Cochran SP; McKinley MP, Masiarz FR. Gel electrophoresis and glass permeation chromatography of the hamster scrapie agent after enzymatic digestion and detergent extraction. *Biochemistry*, 1980, 19, 4892-4898.
[29] Prusiner SB; Hadlow WJ; Eklund CM, Race RE. Sedimentation properties of the scrapie agent. *Proc Natl Acad Sci USA*, 1977, 74, 4656-4660.
[30] Prusiner SB; Hadlow WJ; Eklund CM; Race RE, Cochran SP. Sedimentation characteristics of the scrapie agent from murine spleen and brain. *Biochemistry*, 1978, 17, 4987-4992.
[31] Prusiner SB; Hadlow WJ; Garfin DE; Cochran SP; Baringer JR; Race RE, Eklund CM. Partial purification and evidence for multiple molecular forms of the scrapie agent. *Biochemistry*, 1978, 17, 4993-4997.
[32] Prusiner SB; Bolton DC; Groth DF; Bowman KA; Cochran SP, McKinley MP. Further purification and characterization of scrapie prions. *Biochemistry*, 1982, 21, 6942-6950.
[33] Kellings K; Meyer N; Mirenda C; Prusiner SB, Riesner D. Further analysis of nucleic acids in purified scrapie prion preparations by improved return refocussing gel electrophoresis (RRGE). *J Gen Virol*, 1992, 73, 1025-1029.
[34] Riesner D. Biochemistry and structure of PrP^{C} and PrP^{Sc}. *Br Med Bull*, 2003, 66, 21-33.
[35] Bolton DC; McKinley MP, Prusiner SB. Identification of a protein that purifies with the scrapie prion. *Science*, 1982, 218, 1309-1311.
[36] McKinley MP; Bolton DC, Prusiner SB. A protease-resistant protein is a structural component of the scrapie prion. *Cell*, 1983, 35, 57-62.
[37] Bendheim PE; Barry RA; DeArmond SJ; Stites DP, Prusiner SB. Antibodies to a scrapie prion protein. *Nature*, 1984, 310, 418-421.
[38] Oesch B; Westaway D; Walchli M; McKinley MP; Kent SBH; Aebersold R; Barry RA; Tempst P; Teplow DB; Hood LE; Prusiner SB, Weissmann C. A cellular gene encodes scrapie PrP 27-30 protein. *Cell*, 1985, 40, 735-746.

[39] Caughey B, Raymond GJ. The scrapie-associated form of PrP is made from a cell surface precursor that is both protease- and phospholipase-sensitive. *J Biol Chem*, 1991, 266, 18217-18223.

[40] Prusiner SB; Groth DF; Bolton DC; Kent SB, Hood LE. Purification and structural studies of a major scrapie prion protein. *Cell*, 1984, 38, 127-134.

[41] Chesebro B; Race R; Wehrly K; Nishio J; Bloom M; Lechner D; Bergstrom S; Robbins K; Mayer L; Keith JM; Garon C, Haase A. Identification of scrapie prion protein-specific mRNA in scrapie-infected and uninfected brain. *Nature*, 1985, 315, 331-333.

[42] Westaway D, Prusiner SB. Conservation of the cellular gene encoding the scrapie prion protein. *Nucl Acids Res*, 1986, 14, 2035-2044.

[43] Basler K; Oesch B; Scott M; Westaway D; Walchli M; Groth DF; McKinley MP; Prusiner SB, Weissmann C. Scrapie and cellular PrP isoforms are encoded by the same chromosomal gene. *Cell*, 1986, 46, 417-428.

[44] Westaway D; Goodman PA; Mirenda CA; McKinley MP; Carlson GA, Prusiner SB. Distinct prion proteins in short and long scrapie incubation period mice. *Cell*, 1987, 51, 651-662.

[45] Manson JC; Clarke AR; Hooper ML; Aitchison L; McConnell I, Hope J. 129/Ola mice carrying a null mutation in PrP that abolishes mRNA production are developmentally normal. *Mol Neurobiol*, 1994, 8, 121-127.

[46] Bendheim PE; Brown HR; Rudelli RD; Scala LJ; Goller NL; Wen GY; Kascsak RJ; Cashman NR, Bolton DC. Nearly ubiquitous tissue distribution of the scrapie agent precursor protein. *Neurology*, 1992, 42, 149-156.

[47] Wickner RB; Liebman SW, Saupe SJ. *Prions of yeast and filamentous fungi: [URE3], [PSI$^+$], [PIN$^+$], and [Het-s]*. In: Prusiner SB, editor. Prion Biology and Diseases. 2nd ed. Cold Spring Harbor: Cold Spring Harbor Laboratory Press; 2004. p. 305-372.

[48] Manuelidis L; Sklaviadis T; Akowitz A, Fritch W. Viral particles are required for infection in neurodegenerative Creutzfeldt-Jakob disease. *Proc Natl Acad Sci USA*, 1995, 92, 5124-5128.

[49] Manuelidis L; Yu Z-X; Barquero N, Mullins B. Cells infected with scrapie and Creutzfeldt-Jakob disease agents produce intracellular 25-nm virus-like particles. *Proc Natl Acad Sci USA*, 2007, 104, 1965-1970.

[50] Legname G; Baskakov IV; Nguyen H-OB; Riesner D; Cohen FE; DeArmond SJ, Prusiner SB. Synthetic mammalian prions. *Science*, 2004, 305, 673-676.

[51] Kocisko DA; Come JH; Priola SA; Chesebro B; Raymond GJ; Lansbury PT, Jr., Caughey B. Cell-free formation of protease-resistant prion protein. *Nature*, 1994, 370, 471-474.

[52] Saborio GP; Permanne B, Soto C. Sensitive detection of pathological prion protein by cyclic amplification of protein misfolding. *Nature*, 2001, 411, 810-813.

[53] Castilla J; Saa P; Hetz C, Soto C. *In vitro* generation of infectious scrapie prions. *Cell*, 2005, 121, 195-206.

[54] Ryou C, Mays CE. Prion propagation *in vitro*: are we there yet? *Int J Med Sci* 2008, 5, 347-353

[55] Sparkes RS; Simon M; Cohn VH; Fournier REK; Lem J; Klisak I; Heinzmann C; Blatt C; Lucero M; Mohandas T; DeArmond SJ; Westaway D; Prusiner SB, Weiner LP. Assignment of the human and mouse prion protein genes to homologous chromosomes. *Proc Natl Acad Sci USA*, 1986, 83, 7358-7362.

[56] Stahl N; Borchelt DR; Hsiao K, Prusiner SB. Scrapie prion protein contains a phosphatidylinositol glycolipid. *Cell*, 1987, 51, 229-240.

[57] Brown DR; Qin K; Herms JW; Madlung A; Manson J; Strome R; Fraser PE; Kruck T; von Bohlen A; Schulz-Schaeffer W; Giese A; Westaway D, Kretzschmar H. The cellular prion protein binds copper *in vivo*. *Nature*, 1997, 390, 684-687.

[58] Stahl N; Borchelt DR, Prusiner SB. Differential release of cellular and scrapie prion proteins from cellular membranes by phosphatidylinositol-specific phospholipase C. *Biochemistry*, 1990, 29, 5405-5412.

[59] Rudd PM; Wormald MR; Wing DR; Prusiner SB, Dwek RA. Prion glycoprotein: structure, dynamics, and roles for the sugars. *Biochemistry*, 2001, 40, 3759-3766.

[60] Harris DA; Peters PJ; Taraboulos A; Lingappa V; DeArmond SJ, Prusiner SB. *Cell biology of prions*. In: Prusiner SB, editor. Prion Biology and Diseases. 2nd ed. Cold Spring Harbor: Cold Spring Harbor Laboratory Press; 2004. p. 483-544.

[61] Klein TR; Kirsch D; Kaufmann R, Riesner D. Prion rods contain small amounts of two sphingolipids as revealed by thin-layer chromatography and mass spectrometry. *J Biol Chem*, 1998, 379, 655-666.

[62] Caughey B, Lansbury Jr PT. Protofibirils, pores, fibrils, and neurodegeneration: Separating the responsible protein aggregates from the innocent bystanders. *Annu Rev Neurosci*, 2003, 26, 267-298.

[63] Merz PA; Somerville RA; Wisniewski HM, Iqbal K. Abnormal fibrils from scrapie-infected brain. *Acta Neuropathol (Berl)*, 1981, 54, 63-74.

[64] Merz PA; Kascsak R; Rubenstein R; Carp RI, Wisniewski HM. *Variations in SAF from different scrapie agents*. In: Tateishi J, editor. Proceedings of Workshop on Slow Transmissible Diseases. Tokyo: Japanese Ministry of Health and Welfare; 1984. p. 137-145.

[65] Merz PA; Kascsak RJ; Rubenstein R; Carp RI, Wisniewski HM. Antisera to scrapie-associated fibril protein and prion protein decorate scrapie-associated fibrils. *J Virol*, 1987, 61, 42-49.

[66] Merz PA; Somerville RA, Wisniewski HM. Abnormal fibrils in scrapie and senile dementia of Alzheimer type. *Unconventional Viruses and the Central Nervous System*, 1983, 259-281.

[67] Prusiner SB; McKinley MP; Bowman KA; Bolton DC; Bendheim PE; Groth DF, Glenner GG. Scrapie prions aggregate to form amyloid-like birefringent rods. *Cell*, 1983, 35, 349-358.

[68] Bueler H; Fisher M; Lang Y; Bluethmann H; Lipp H-P; DeArmond SJ; Prusiner SB; Aguet M, Weissmann C. Normal development and behaviour of mice lacking the neuronal cell-surface PrP protein. *Nature*, 1992, 356, 577-582.

[69] Moore RC; Lee IY; Silverman GL; Harrison PM; Strome R; Heinrich C; Karunaratne A; Pasternak SH; Chishti MA; Liang Y; Mastrangelo P; Wang K; Smit AFA; Katamine S; Carlson GA; Cohen FE; Prusiner SB; Melton DW; Tremblay P; Hood LE, Westaway D. Ataxia in prion protein (PrP)-deficient mice is associated with upregulation of the novel PrP-like protein doppel. *J Mol Biol*, 1999, 292, 797-817.

[70] Rossi D; Cozzio A; Flechsig E; Klein MA; Rulicke T; Aguzzi A, Weissmann C. Onset of ataxia and Purkinje cell loss in PrP null mice inversely correlated with Dpl level in brain. *EMBO J*, 2001, 20, 694-702.

[71] Sakaguchi S; Katamine S; Nishida N; Moriuchi R; Shigematsu K; Sugimoto T; Nakatani A; Kataoka Y; Houtani T; Shirabe S; Okada H; Hasegawa S; Miyamoto T, Noda T. Loss of cerebellar Purkinje cells in aged mice homozygous for a disrupted PrP gene. *Nature*, 1996, 380, 528-531.

[72] Cashman NR; Loertscher R; Nalbantoglu J; Shaw I; Kascsak RJ; Bolton DC, Bendheim PE. Cellular isoform of the scrapie agent protein participates in lymphocyte activation. *Cell*, 1990, 61, 185-192.

[73] Maglio LE; Perez MF; Martins VR; Brentani RR, Ramirez OA. Hippocampal synaptic plasticity in mice devoid of cellular prion protein. *Mol Brain Res*, 2004, 131, 58-64.

[74] Chiarini LB; Freitas AR; Zanata SM; Brentani RR; Martins VR, Linden R. Cellular prion protein transduces neuroprotective signals. *EMBO J*, 2002, 21, 3317-3326.

[75] Mouillet-Richard S; Ermonval M; Chebassier C; Laplanche J-L; Lehmann S; Launay JM, Kellermann O. Signal transduction through prion protein. *Science*, 2000, 289, 1925-1928.

[76] Brown LR, Harris DA. *The prion protein and copper: What is the connection?* In: Massaro EJ, editor. Handbook of copper pharmacology and toxicology. Totowa: Humana Press; 2002. p. 103-113.

[77] Zhang CC; Steele AD; Lindquist S, Lodish HF. Prion protein is expressed on long-term repopulating hematopoietic stem cells and is important for their self-renewal. *Proc Natl Acad Sci U S A*, 2006, 103, 2184-2189.

[78] Steele AD; Emsley JG; Ozdinler PH; Lindquist S, Macklis JD. Prion protein (PrP^C) positively regulates neural precursor proliferation during developmental and adult mammalian neurogenesis. *Proc Natl Acad Sci U S A*, 2006, 103, 3416-3421.

[79] Fischer M; Rulicke T; Raeber A; Sailer A; Moser M; Oesch B; Brandner S; Aguzzi A, Weissmann C. Prion protein (PrP) with amino-proximal deletions restoring susceptibility of PrP knockout mice to scrapie. *EMBO J*, 1996, 15, 1255-1264.

[80] Telling GC; Haga T; Torchia M; Tremblay P; DeArmond SJ, Prusiner SB. Interactions between wild-type and mutant prion proteins modulate neurodegeneration in transgenic mice. *Genes Dev*, 1996, 10, 1736-1750.

[81] Mays CE; Titlow W; Seward T; Telling GC, Ryou C. Enhancement of protein misfolding cyclic amplification by using concentrated cellular prion protein source. *Biochem Biophys Res Commun*, 2009, 388, 306-310.

[82] Bueler H; Aguzzi A; Sailer A; Greiner R-A; Autenried P; Aguet M, Weissmann C. Mice devoid of PrP are resistant to scrapie. *Cell*, 1993, 73, 1339-1347.

[83] Manson JC; Clarke AR; McBride PA; McConnell I, Hope J. PrP gene dosage determines the timing but not the final intensity or distribution of lesions in scrapie pathology. *Neurodegeneration*, 1994, 3, 331-340.

[84] Prusiner SB; Groth D; Serban A; Koehler R; Foster D; Torchia M; Burton D; Yang S-L, DeArmond SJ. Ablation of the prion protein (PrP) gene in mice prevents scrapie and facilitates production of anti-PrP antibodies. *Proc Natl Acad Sci USA*, 1993, 90, 10608-10612.

[85] Carlson GA; Kingsbury DT; Goodman PA; Coleman S; Marshall ST; DeArmond S; Westaway D, Prusiner SB. Linkage of prion protein and scrapie incubation time genes. *Cell*, 1986, 46, 503-511.

[86] Carlson GA; Lovett M; Epstein CJ; Westaway D; Goodman PA; Marshall ST, Prusiner SB. The prion gene complex: polymorphism of *Prn-p* and its linkage with agouti. *XIVth Molecular and Biochemical Genetics Workshop, Bar Harbor, Maine*, 1986,

[87] Carlson GA; Westaway D; DeArmond SJ; Peterson-Torchia M, Prusiner SB. Primary structure of prion protein may modify scrapie isolate properties. *Proc Natl Acad Sci USA*, 1989, 86, 7475-7479.

[88] Carlson GA; Westaway D; Goodman PA; Peterson M; Marshall ST, Prusiner SB. *Genetic control of prion incubation period in mice*. In: Bock G, Marsh J, editors. Novel Infectious Agents and the Central Nervous System, Ciba Foundation Symposium 135. Chichester, UK: John Wiley and Sons; 1988. p. 84-99.

[89] Dickinson AG; Meikle VMH, Fraser H. Identification of a gene which controls the incubation period of some strains of scrapie agent in mice. *J Comp Pathol*, 1968, 78, 293-299.

[90] Carp RI; Moretz RC; Natelli M, Dickinson AG. Genetic control of scrapie: incubation period and plaque formation in □mice. *J Gen Virol*, 1987, 68, 401-407.

[91] Moore RC; Hope J; McBride PA; McConnell I; Selfridge J; Melton DW, Manson JC. Mice with gene targetted prion protein alterations show that *Prnp*, *Sinc* and *Prni* are congruent. *Nat Genet*, 1998, 18, 118-125.

[92] Goldmann W; Hunter N; Benson G; Foster JD, Hope J. Different scrapie-associated fibril proteins (PrP) are encoded by lines of sheep selected for different alleles of the *Sip* gene. *J Gen Virol*, 1991, 72, 2411-2417.

[93] Laplanche J-L; Chatelain J; Westaway D; Thomas S; Dussaucy M; Brugere-Picoux J, Launay J-M. PrP polymorphisms associated with natural scrapie discovered by denaturing gradient gel electrophoresis. *Genomics*, 1993, 15, 30-37.

[94] Collinge J; Beck J; Campbell T; Estibeiro K, Will RG. Prion protein gene analysis in new variant cases of Creutzfeldt-Jakob disease. *Lancet*, 1996, 348, 56.

[95] Collinge J; Whitfield J; McKintosh E; Beck J; Mead S; Thomas DJ, Alpers MP. Kuru in the 21st century--an acquired human prion disease with very long incubation periods. *Lancet*, 2006, 367, 2068-2074.

[96] Pan K-M; Baldwin M; Nguyen J; Gasset M; Serban A; Groth D; Mehlhorn I; Huang Z; Fletterick RJ; Cohen FE, Prusiner SB. Conversion of α-helices into β-sheets features in the formation of the scrapie prion proteins. *Proc Natl Acad Sci USA*, 1993, 90, 10962-10966.

[97] Safar J; Roller PP; Gajdusek DC, Gibbs CJ, Jr. Conformational transitions, dissociation, and unfolding of scrapie amyloid (prion) protein. *J Biol Chem*, 1993, 268, 20276-20284.

[98] Caughey BW; Dong A; Bhat KS; Ernst D; Hayes SF, Caughey WS. Secondary structure analysis of the scrapie-associated protein PrP 27-30 in water by infrared spectroscopy. *Biochemistry*, 1991, 30, 7672-7680.

[99] Riek R; Hornemann S; Wider G; Billeter M; Glockshuber R, Wuthrich K. NMR structure of the mouse prion protein domain PrP(121-231). *Nature*, 1996, 382, 180-182.

[100] Riek R; Hornemann S; Wider G; Glockshuber R, Wuthrich K. NMR characterization of the full-length recombinant murine prion protein, *m*PrP(23-231). *FEBS Lett*, 1997, 413, 282-288.

[101] James TL; Liu H; Ulyanov NB; Farr-Jones S; Zhang H; Donne DG; Kaneko K; Groth D; Mehlhorn I; Prusiner SB, Cohen FE. Solution structure of a 142-residue recombinant

prion protein corresponding to the infectious fragment of the scrapie isoform. *Proc Natl Acad Sci USA*, 1997, 94, 10086-10091.

[102] Burns CS; Aronoff-Spencer E; Legname G; Prusiner SB; Antholine WE; Gerfen GJ; Peisach J, Millhauser GL. Copper coordination in the full-length, recombinant prion protein. *Biochemistry*, 2003, 42, 6794-6803.

[103] Billeter M; Riek R; Wider G; Hornemann S; Glockshuber R, Wuthrich K. Prion protein NMR structure and species barrier for prion diseases. *Proc Natl Acad Sci USA*, 1997, 94, 7281-7285.

[104] Christen B; Hornemann S; Damberger FF, Wüthrich K. Prion protein NMR structure from tammar wallaby (Macropus eugenii) shows that the b2-a2 loop is modulated by long-range sequence effects. *J Mol Biol*, 2009, 389, 833-845.

[105] Christen B; Pérez DR; Hornemann S, Wüthrich K. NMR structure of the bank vole prion protein at 20 °C contains a structured loop of residues 165-171. *J Mol Biol* 2008, 383, 306-312.

[106] Liu H; Farr-Jones S; Ulyanov NB; Llinas M; Marqusee S; Groth D; Cohen FE; Prusiner SB, James TL. Solution structure of Syrian hamster prion protein rPrP(90-231). *Biochemistry*, 1999, 38, 5362-5377.

[107] Gossert AD; Bonjour S; Lysek DA; Fiorito F, Wuthrich K. Prion protein NMR structures of elk and of mouse/elk hybrids. *Proc Natl Acad Sci USA*, 2005, 102, 646-650.

[108] Calzolai L; Lysek DA; Guntert P; von Schroetter C; Riek R; Zahn R, Wuthrich K. NMR structures of three single-residue variants of the human prion protein. *Proc Natl Acad Sci USA*, 2000, 97, 8340-8345.

[109] Lopez-Garcia F; Zahn R; Riek R, Wuthrich K. NMR structure of the bovine prion protein. *Proc Natl Acad Sci USA*, 2000, 97, 8334-8339.

[110] Zahn R; Liu A; Luhrs T; Riek R; von Schroetter C; Lopez-Garcia F; Billeter M; Calzolai L; Wider G, Wuthrich K. NMR solution structure of the human prion protein. *Proc Natl Acad Sci USA*, 2000, 97, 145-150.

[111] Lysek DA; Schorn C; Nivon LG; Esteve-Moya V; Christen B; Calzolai L; von Schroetter C; Fiorito F; Herrmann T; Guntert P, Wuthrich K. Prion protein NMR structures of cats, dogs, pigs, and sheep. *Proc Natl Acad Sci USA*, 2005, 102, 640-645.

[112] Calzolai L; Lysek DA; Pérez DR; Guntert P, Wuthrich K. Prion protein NMR structures of chickens, turtles, and frogs. *Proc Natl Acad Sci USA*, 2005, 102, 651-655.

[113] Haire LF; Whyte SM; Vasisht N; Gill AC; Verma C; Dodson EJ; Dodson GG, Bayley PM. The crystal structure of the globular domain of sheep prion protein. *J Mol Biol*, 2004, 336, 1175-1183.

[114] Kanyo ZF; Pan K-M; Williamson A; Burton DR; Prusiner SB; Fletterick RJ, Cohen FE. Antibody binding defines a structure for an epitope that participates in the PrP^{C} --> PrP^{Sc} conformational change. *J Mol Biol*, 1999, 293, 855-863.

[115] Eghiaian F; Grosclaude J; Lesceu S; Debey P; Doublet B; Treguer E; Rezaei H, Knossow M. Insight into the PrP^{C}-->PrP^{Sc} conversion from the structures of antibody-bound ovine prion scrapie-susceptibility variants. *Proc Natl Acad Sci U S A*, 2004, 101, 10254-10259.

[116] Meyer RK; Lustig A; Oesch B; Fatzer R; Zurbriggen A, Vandevelde M. A monomer-dimer equilibrium of a cellular prion protein (PrP^{C}) not observed with recombinant PrP. *J Biol Chem*, 2000, 275, 38081-38087.

[117] Knaus KJ; Morillas M; Swietnicki W; Malone M; Surewicz WK, Yee VC. Crystal structure of the human prion protein reveals a mechanism for oligomerization. *Nat Struct Biol*, 2001, 8, 770-774.

[118] Wille H; Michelitsch MD; Guenebaut V; Supattapone S; Serban A; Cohen FE; Agard DA, Prusiner SB. Structural studies of the scrapie prion protein by electron crystallography. *Proc Natl Acad Sci USA*, 2002, 99, 3563-3568.

[119] Govaerts C; Wille H; Prusiner SB, Cohen FE. Evidence for assembly of prions with left-handed β-helices into trimers. *Proc Natl Acad Sci USA*, 2004, 101, 8342-8347.

[120] DeMarco ML, Daggett V. From conversion to aggregation: protofibril formation of the prion protein. *Proc Natl Acad Sci U S A*, 2004, 101, 2293-2298.

[121] DeMarco ML; Silveira J; Caughey B, Daggett V. Structural properties of prion protein protofibrils and fibrils: an experimental assessment of atomic models *Biochemistry*, 2006, 45, 15573-15582.

[122] Lu X; Wintrode PL, Surewicz WK. β-sheet core of human prion protein amyloid fibrils as determined by hydrogen/deuterium exchange. *Proc Natl Acad Sci USA*, 2007, 104, 1510-1515.

[123] Cobb NJ; Sonnichsen FD; Mchaourab H, Surewicz WK. Molecular architecture of human prion protein amyloid: A parallel, in-register b-structure. *Proc Natl Acad Sci USA*, 2007, 104, 18946-18951.

[124] Caughey B; Baron GS; Chesebro B, Jeffrey M. Getting a grip on prions: oligomers, amyloids, and pathological membrane interactions. *Annu Rev Biochem*, 2009, 78, 177-204.

[125] Jarrett JT, Lansbury PT, Jr. Seeding "one-dimensional crystallization" of amyloid: a pathogenic mechanism in Alzheimer's disease and scrapie? *Cell*, 1993, 73, 1055-1058.

[126] Prusiner SB. Molecular biology of prion diseases. *Science*, 1991, 252, 1515-1522.

[127] Kaneko K; Zulianello L; Scott M; Cooper CM; Wallace AC; James TL; Cohen FE, Prusiner SB. Evidence for protein X binding to a discontinuous epitope on the cellular prion protein during scrapie prion propagation. *Proc Natl Acad Sci USA*, 1997, 94, 10069-10074.

[128] Telling GC; Scott M; Mastrianni J; Gabizon R; Torchia M; Cohen FE; DeArmond SJ, Prusiner SB. Prion propagation in mice expressing human and chimeric PrP transgenes implicates the interaction of cellular PrP with another protein. *Cell*, 1995, 83, 79-90.

[129] Cohen FE, Prusiner SB. Pathologic conformations of prion proteins. *Annu Rev Biochem*, 1998, 67, 793-819.

[130] Makarava N, Baskakov IV. The same primary structure of the prion protein yields two distinct self-propagating states. *J Biol Chem*, 2008, 283, 15988-15996.

[131] Wille H; Shanmugam M; Murugesu M; Ollesch J; Stubbs G; Long JR; Safar JG, Prusiner SB. Surface charge of polyoxometalates modulates polymerization of the scrapie prion protein. *Proc Natl Acad Sci USA*, 2009, 106, 3740-3745.

[132] Calzolai L, Zahn R. Influence of pH on NMR structure and stability of the human prion protein globular domain. *J Biol Chem*, 2003, 278, 35592-35596.

[133] Gambetti P; Kong Q; Zou W; Parchi P, Chen SG. Sporadic and familial CJD: classification and characterisation. *Br Med Bull*, 2003, 66, 213-239.

[134] Lasmezas CI; Deslys J-P; Robain O; Jaegly A; Beringue V; Peyrin J-M; Fournier J-G; Hauw J-J; Rossier J, Dormont D. Transmission of the BSE agent to mice in the absence of detectable abnormal prion protein. *Science*, 1997, 275, 402-405.

[135] Hill AF; Joiner S; Linehan J; Desbruslais M; Lantos PL, Collinge J. Species-barrier-independent prion replication in apparently resistant species. *Proc Natl Acad Sci USA*, 2000, 97, 10248-10253.

[136] Deleault NR; Lucassen RW, Supattapone S. RNA molecules stimulate prion protein conversion. *Nature*, 2003, 425, 717-720.

[137] Deleault NR; Geoghegan JC; Nishina K; Kascsak R; Williamson RA, Supattapone S. Protease-resistant prion protein amplification reconstituted with partially purified substrates and synthetic polyanions. *J Biol Chem*, 2005, 280, 26873-26879.

[138] Sayers EW; Barrett T; Benson DA; Bryant SH; Canese K; Chetvernin V; Church DM; DiCuccio M; Edgar R; Federhen S; Feolo M; Geer LY; Helmberg W; Kapustin Y; Landsman D; Lipman DJ; Madden TL; Maglott DR; Miller V; Mizrachi I; Ostell J; Pruitt KD; Schuler GD; Sequeira E; Sherry ST; Shumway M; Sirotkin K; Souvorov A; Starchenko G; Tatusova TA; Wagner L; Yaschenko E; Ye J. Database resources of the National Center for Biotechnology Information. *Nucl Acids Res*, 2009, D5-15.

In: Molecular Aspects of Infectious Diseases
Editor: Mazen T. Saleh ISBN: 978-1-61728-690-2

Chapter IX

Transmissible Spongiform Encephalopathy (TSE)

Chongsuk Ryou *

Sanders-Brown Center on Aging, Department of Microbiology, Immunology and Molecular Genetics, College of Medicine, University of Kentucky, 800 Rose St. HSRB-326, Lexington, KY 40536, U.S.A.

Abstract

Transmissible spongiform encephalopathies (TSEs), also known as prion diseases, are infectious neurodegenerative disorders caused by prions, the unconventional agents composed of proteins but devoid of nucleic acids. TSEs are transmitted to several mammalian species including humans. As shown in the outbreaks of bovine spongiform encephalopathy (BSE) and subsequent transmission of BSE to humans causing variant Creutzfeldt-Jakob disease, TSEs pose a potential risk for human public health and may result in many unprecedented problems. Although much remains enigmatic, here, we discuss TSEs of individual host species with the focus on transmission of disease. In this chapter, we describe the current understanding on the species barrier and prion strains in the context of intra- and inter-species transmission. We also address prion pathogenesis and the factors influencing it. In the discussion of human and animals TSEs, we review individual TSEs with the details on etiology, clinical and pathological characterization, the mode of transmission, and the emergence of atypical TSEs.

Introduction

Transmissible spongiform encephalopathies (TSEs), also known as prion diseases, are progressive fatal neurodegenerative diseases of humans and other mammalian species [1]. TSEs have captured much attention since the appearance of bovine spongiform

* Corresponding author: Tel (859) 257 4016, Fax (859) 257 8382, E-mail cryou2@email.uky.edu

encephalopathy (BSE, also known as "mad cow disease") and transmission of BSE to humans in the United Kingdom and European countries. Because of its unprecedented transmission and devastating outcomes, BSE has influenced medical, agricultural, economic, and political issues in our society. BSE has been found in more than 25 countries worldwide and has spread to non-European countries including the United States, Canada, and Japan [2]. In addition, chronic wasting disease (CWD), a type of TSE in cervids of North America, has occurred in the Rocky Mountain area and spread to other states and provinces of the United States and Canada. Although there is no strong evidence supporting that CWD has been transmitted to humans, the high incidence of CWD raises concerns for a potential risk to humans. Because the wild ruminants affected by CWD are found in the area where cattle and sheep graze, it is possible to speculate that CWD could be transmitted to domestic animals and cross the species barrier for human infection similar to BSE. Despite the fact that TSEs are potential risks for humans and animals, there is no therapy and early diagnosis available.

The causative agent of TSEs is an unconventional proteinaceous particle termed prion [1, 3]. These prion agents are devoid of functional DNA and/or RNA that transfer genetic information. Despite lacking genetic material, prions replicate through self-perpetuating conformational conversion of cellular prion protein (PrP^{C}) to pathogenic, scrapie-associated prion protein (PrP^{Sc}), which is a major, if not sole, component of prions (See Chapter 5 for details). This biochemical event is the hallmark of TSEs and distinguishes TSEs from other amyloidogenic neurodegenerative diseases. The ability of prions to adopt an abnormal conformation confers the infectious nature of pathogens and transmissibility of disease.

Transmission of TSEs

Species Barrier

Transmission of TSEs is relatively efficient within the same species [4, 5]. Intra-species transmission usually occurs after a short incubation period with a complete attack rate. In contrast, transmission of TSEs between species is much less efficient [4, 5]. Only in limited cases, inter-species transmission is possible after an extended incubation time with a low attack rate. For instance, in a laboratory setting, mouse prions readily infect mice, but not hamsters or transgenic mice expressing hamster prion protein (PrP) [6, 7]. This phenomenon involved in inefficient transmission of a TSE agent to a different species is referred to as the species barrier. As learned from transmission of natural prion agents to rodents, the species barrier is abrogated by multiple passages of a TSE agent to the same host species [8]. It appears that adaptation of the agent occurs during this period. Under natural conditions, the species barrier prevents the transmission of heterologous prions to a host species. However, the onset of the disease is not limited to a certain species. TSEs can be transmissible from one host species to another by overcoming the species barrier. The recent emergence of the new variant Creutzfeldt-Jakob disease (vCJD) in humans has highlighted the ability of prions to cross the species barrier and propagate in other species. In the cases of vCJD, humans might become infected with prions after ingesting meat from cattle infected by BSE [9].

Details on TSE transmission events are poorly understood. We currently know that the major contributors influencing transmission of TSEs are PrP^C of hosts, PrP^{Sc} of TSE agents, and possibly, interplay between these two factors. TSEs are associated with expression of the prion protein (*PRNP*) gene [10-12]. The availability of PrP^C directly influences PrP^{Sc} propagation that is closely correlated with transmission of disease. *PRNP* deficient ($Prnp^{o/o}$) mice become completely resistant to disease with no trace of clinical symptoms or neuropathology [13-15], whereas transgenic mouse lines expressing an excess level of PrP^C can accelerate the disease process as evidenced by abbreviated incubation periods [16, 17]. Furthermore, inter-species transmission studies suggested that the difference in PrP primary structure of the host and donor species determines the species barrier and controls transmission of TSEs. For instance, hamster prions inefficient to be transmitted to mice readily spread the disease in transgenic mice expressing hamster PrP^C [6, 18]. Similarly, transmission of disease in mice by CWD prions is not efficient [19]. However, the transgenic lines expressing cervid PrP^C are efficiently affected by the same prions [20]. In intra-species transmission, natural sequence variations in PrP^C control transmission of TSE. Susceptibilities of scrapie vary depending on the different *PRNP* genotypes of sheep [21, 22]. Polymorphisms at codons 136, 154 and 171 of the ovine *PRNP* gene are particularly important [23]. The VRQ homozygous alleles at these positions are highly susceptible to scrapie whereas the ARR homozygous alleles exhibit resistance [24, 25]. Thus, minimal divergences found in the polymorphic codons of PrP^C are critical in determining efficient transmission of TSEs.

Prion Strain

Prions have strains that are maintained through serial transmissions. Individual prion strains manifest distinct disease phenotypes. Different prion strains affecting the same host species are characterized by difference in incubation periods, the pattern of histopathologic lesions in the brain, distribution of prion deposits in the brain, organ tropism, and clinical symptoms in vivo [4, 5]. These specific phenotypic traits are likely to be encoded by PrP^{Sc}, but not by differences in the PrP^C sequence, because the propagation of different prion strains in a host relies on PrP^C of the identical *PRNP* gene. It has been suggested that prion strain specificity might be enciphered within conformational differences in the tertiary structure of PrP^{Sc} [26]. Prions lack great diversity in amino acid sequences across the different strains because the PrP sequence is highly conserved among mammalian species. Nonetheless, the diversity of prion strains is believed to be determined by the divergent conformation of PrP^{Sc}. Indeed, recent studies demonstrated the existence of variable PrP^{Sc} conformations [27, 28], suggesting the conformational flexibility of PrP^{Sc}. Because diversity of prion strains appears to be determined by PrP^{Sc} species that differ in their conformations, prion strains are also distinguishable by different biochemical parameters such as glycosylated patterns, stability against denaturing agents, and the electrophoretic mobility of protease resistant abnormal PrP^{Sc} [5, 29, 30]. The strain-specific differences in the molecular sizes of proteinase resistant PrP^{Sc} reflect a different access of proteinase K to the N-terminal region of the PrP molecule according to individually different conformations.

Prion Pathogenesis

Prion pathogenesis is not completely understood and much remains elusive. However, it is clear that prion replication is central in the entire pathogenesis because prions are responsible for causing TSEs [31]. Prion replication is involved in the conversion of the host-encoded PrP^{C} to the abnormally folded PrP^{Sc} [1]. Because PrP^{Sc} becomes hydrophobic after conformational changes, it tends to form aggregates or, occasionally, prion amyloids [32-34]. The PrP^{Sc} aggregates are thought to be toxic to neurons and drive neuronal cell death, which ultimately results in neurodegeneration [35, 36]. In addition to the close correlation of PrP^{Sc} accumulation with the progression of disease and disease-specific pathologic changes in vivo [37], highly purified PrP^{Sc} can cause harmful effects directly to cultured neuronal cells [38, 39]. One the other hand, some studies suggest that PrP^{Sc} can cause pathologic effects indirectly through glial cells by triggering signal transduction cascades for neuronal apoptosis [38, 39]. Other lines of studies have suggested that soluble oligomers of PrP^{Sc}, not PrP^{Sc} aggregates or amyloids, are toxic species causing the impairment of cellular functions [40, 41]. Although prion pathogenesis is associated with formation of PrP^{Sc} aggregates and their accumulation, it is not clear whether the gain of toxic PrP^{Sc} function is responsible for the downstream events toward neurotoxicity, cellular stress, cell death, and neurodegeneration. An alternatively proposed mechanism for prion pathogenesis is the loss of physiological PrP^{C} functions. Due to the conformational conversion of PrP^{C} to PrP^{Sc}, the physiological role of PrP^{C} for cellular survival or protection against insulting signals can be disrupted [42, 43].

In addition to the influence of PrP molecules, prion pathogenesis appears to be associated with other host components. Although the central nervous system (CNS) is the primary target for prion pathogenesis, the lymphoid system plays an important role in the early stages of prion infection [44]. When the hosts are exposed to prions by oral routes, the ingested prions are taken up through the gastrointestinal tract by the function of M cells (intestinal epithelial cells that transfer antigens including pathogens through the epithelium) in the Peyer's patches [45]. Once migrated, prions primarily propagate and are accumulated in the secondary lymphoid organs and tissues in periphery such as lymph nods and spleen long before PrP^{Sc} appears in the brain [44, 46, 47]. Later, prions in the affected peripheral organs and tissues invade the CNS along with the migration through the peripheral nerves [48]. Although the lymphoreticular system does not appear to be essential for neuroinvasion, the roles of follicular dendritic cells and B lymphocytes are critical in this series of events [49-52].

TSEs in Humans and Animals

For many years, TSEs have occurred in several mammalian species including humans. A prototypic TSE, scrapie in sheep, was documented many centuries ago. In contrast, new TSEs such as BSE and vCJD have emerged recently. The mode of transmission for TSEs differs in individual diseases of a certain host species although transmissibility represents one of the most unique features of diseases belonging to this group. The most notable TSEs in humans and animals include Creutzfeldt-Jakob disease (CJD) in humans, BSE in cattle, scrapie in sheep, and CWD in deer and elks.

Clinical and Pathological Features

TSEs manifest the neurological conditions characterized by progressive, but invariably fatal, degeneration in the CNS during long incubation periods [1] that range from months to years and, in the case of some kuru patients, may have been as long as 40 years [53]. Human and animal hosts with TSEs manifest a number of clinical features involved in impaired brain function [54]. The most distinctive clinical signs of human patients with TSEs are ataxia and dementia. Animals affected by TSEs exhibit neurological symptoms such as loss of coordination, gait disturbance, irritable demeanor, and severe wasting. Neuropathologically, TSEs feature activated astrocytes and the presence of large vacuoles that generate a spongiform appearance in the brain [55]. Additionally, the most characteristic pathological presentation is the accumulation of scrapie-associated fibrils or prion rods consisting of abnormal protease-resistant PrP^{Sc}. Unlike the pathology caused by other conventional pathogens, the hosts affected by TSEs lack immune responses to prions such as generation of anti-prion antibody and inflammatory responses to pathogens [44]. Presumably, they are not recognized as foreign by the immune system.

Human TSEs

Since the first cases of CJD were reported in the 1920s [56, 57], Gerstmann-Sträussler-Scheinker syndrome (GSS) [58], a genetic form of human TSEs, and kuru [59], an infectious form of human TSEs, were first described in 1930s and 1950s. Although the etiology of some human TSEs is unknown, it obviously varies for individual TSEs. The transmissible pathogens can be available from either external or internal sources. Depending on how the disease was initiated or transmitted by these pathogens, human TSEs are categorized into three groups: idiopathic, inherited, and acquired. Human TSEs are summarized in Table 1.

Table 9.1. Classification of human TSEs.

Type of TSEs	Comments
Idiopathic	Cause unknown, sporadic occurrence of diseases, no exposure to infectious prion agents sCJD; sFI
Inherited	Associated with pathogenic germ-line mutation of the *PRNP* gene; no exposure to infectious prion agents fCJD; GSS; FFI
Acquired	Exposure to infectious prion agents; ritualistic cannibalism; exposure to contaminated surgical equipment or tissue transplants (dura mater, cornea, blood, hormone); ingestion of zoonotic BSE prions contaminated in food kuru; iCJD; vCJD

sCJD (sporadic Creutzfeldt-Jakob disease), sFI (sporadic fatal insomnia), fCJD (familial Creutzfeldt-Jakob disease), GSS (Gersmann-Straussler-Scheinker syndrome), FFI (fatal familial insomnia), iCJD (iatrogenic Creutzfeldt-Jakob disease), vCJD (variant Creutzfeldt-Jakob disease)

Idiopathic and inherited forms of TSEs are initiated by internal causes such as a spontaneous change in the conformation of PrP^{C} to PrP^{Sc} and somatic or germ line mutations on the *PRNP* gene. The cause of sporadic (s) CJD is unknown, but it is believed to be developed from random spontaneous protein-misfolding of PrP^{C} into PrP^{Sc}. When there are mutations present in the *PRNP* gene, the mutated PrP^{C} is more prone to be refolded into disease-associated PrP^{Sc} than wild type PrP^{C}, which leads to development of inherited forms of human TSEs including familial (f) CJD, GSS, and fatal familial insomnia (FFI). The acquired forms of TSEs are caused by infection through ingestion or exposure to the external prion material derived from either a homologous or a heterologous source. Accidental exposure to prions during surgical operation or medical treatment causes iatrogenic (i) CJD. Ingestion of material contaminated by infected tissues with externally derived prions leads to development of vCJD and kuru.

Idiopathic TSEs: sCJD and Sporadic Fatal Insomnia

sCJD is the most common human TSE. Most CJD patients (~85%) represent sCJD [60]. The incidence rate of sCJD is 1-2 million individuals per year. The age of onset is between 50 to 70 years. The main symptoms are dementia and rapid decline in cognitive function presenting as confusion, memory loss, and bizarre behavior. Ataxia and myoclonus caused by cerebellar dysfunction are also frequently presented as clinical signs. After the first clinical symptoms are observed, death usually occurs within 1 to 12 months. Pathologically, astrocytosis and spongiform changes associated with neuronal loss are in present in gray matter. Protease resistant PrP^{Sc} is detected in the brain of sCJD patients, while PrP amyloid plaques are found in about 5% of cases. Although acquisition of a TSE agent in sCJD patients remains a mystery, spontaneous formation of PrP^{Sc} without involvement of germ line or somatic cell mutations in the *PRNP* gene of individual patients is hypothesized to be a cause of sCJD.

Sporadic fatal insomnia [61] is an extremely rare form of human TSEs and is not associated with mutations in the *PRNP* gene. However, the *PRNP*-129 genotype has been proposed to be associated with a certain clinical phenotype of sporadic fatal insomnia. The clinical features of this disease include prominent sleep and autonomic disturbances. In rare cases, a slowly progressive dementia is also recognized.

Inherited TSEs: fCJD, GSS and FFI

Inherited forms of TSEs represent nearly 15% of entire human TSE cases and include fCJD, GSS, and FFI [62]. These diseases are defined as progressive neuropsychiatric syndromes caused by genetic mutations of the *PRNP* gene (Table 2). The inheritance is autosomal dominant with high penetrance. The primary clinical symptoms are ataxia, dementia or sleep disturbance. The age of onset and clinical and pathological presentations are variable depending on the particular mutations. fCJD occurs with point mutations found in the region of the second and the third helices of the carboxyl-terminus of PrP^{C}. GSS occurs with point mutations found in the same regions, as well as the hydrophobic region located between octarepeats and the first helix of PrP^{C}. Furthermore, insertions in the octarepeat region at the amino-terminus and a premature termination at codons 145 and 160 are also associated with inherited human TSEs. Insertion of additional 2-9 octapeptides in the octarepeat regions have been detected in various families. Additionally, the variable disease phenotypes appear to be influenced by the *PRNP*-129 genotype. The D178N mutation associated with 129V, for instance, presents with clinical symptoms of CJD, whereas the same mutation associated with 129M presents with the clinicopathological entity of FFI. Because individual patients of the same family show clinical variability, it has been proposed that as yet unidentified modifiers or cofactors such as exogenous viral agents, presence of genes other than *PRNP,* and non-genetic dietary or environmental factors may additionally influence these diseases.

Table 9.2. Mutations of the PRNP gene causing inherited human TSEs.

Diseases	Disease-causing mutations
fCJD	Octarepeat insertions, D178N-129V, V180I, T188K, T188K-129V, E196K, E200K, V203I, E208H, V210I, E211Q, M232R
GSS	Octarepeat insertions, P102L-129M, P105L-129V, A117V-129V, G131-129M, Y145Stop-129M, H187R-129V, F189S-129V, D202N-129V, Q212P, Q217R-129M
FFI	D178N-129M
Not classified	G114V, Q160Stop-129M, N171S, T183A, H187R

Acquired TSEs: kuru, iCJD, and vCJD

This group of human TSEs includes kuru, iCJD, and vCJD [54]. Exposure to TSE agents by contact with or consumption of brain tissues or extracts contaminated by TSE agents leads to the development of this group of diseases. Kuru was first noted as an endemic disease in the Fore tribes of the Eastern Highlands Province of Papua New Guinea in the 1950s [59]. As indicated in the term kuru, taken from the Fore word "kuria/guria"("to shake"), the clinical hallmark of the disease is shivering due to cerebellar ataxia. The ritualistic cannibalism practiced in the funeral of deceased relatives who were affected by kuru is attributed to the route of transmission. The origin of the outbreak is considered to be a deceased individual with sporadic human TSE. The incidence of kuru has decreased after banning cannibalism [53].

iCJD is transmitted during surgical procedures or medical treatment. Clinical signs of iCJD vary depending on TSE agents transmitted to patients. Iatrogenic transmission of TSE agents was first described in patients who received corneal transplantation of a graft derived from a sCJD patient [63]. iCJD was also found in the individuals who received dura mater implants obtained from cadavers afflicted with TSEs [64, 65]. Treatment with human growth hormone extracted from pituitary glands pooled from large groups of individuals caused iCJD in the growth hormone recipients [66]. Apparently, the hormones were contaminated by TSE agents from the brain tissue of individuals who had CJD but were not diagnosed. iCJD can be transmitted during neurosurgical procedures using instruments contaminated with TSE agents [67]. Stereotactic placement of electrodes incompletely sterilized following use on individuals with CJD developed iCJD. More recently, a few cases of vCJD transmission by blood transfusion have been reported in the United Kingdom [68-70]. These alerting reports raise additional human public health concerns because TSEs can be transmitted preclinically by infected carriers of vCJD.

In the mid-1990s, vCJD was described first in the United Kingdom as a new variant form of CJD in humans [9]. vCJD is zoonotically transmitted to humans from BSE by dietary consumption of beef or meat products contaminated with BSE agents. Since 1995, approximately 200 individuals have succumbed from vCJD worldwide, mostly limited to the United Kingdom (Table 3 and Figure 9.1). Fortunately since 2001, the incidence of this disease has decreased to a reported rate of less than 10 cases per year worldwide. There is no evidence that vCJD is associated with mutations with the *PRNP* gene. However, it appears that association with the polymorphic *PRNP* 129 genotype is important for transmission of BSE to humans [71]. All vCJD patients showed the MM genotype at this codon. Unlike classic senile CJD, vCJD affected young individuals in their 20s and early 30s. In addition to the early age of onset, distinct clinical and neuropathological features distinguished vCJD from sCJD. Clinical presentation of vCJD includes psychiatric symptoms such as behavioral changes, anxiety, depression, and withdrawal. vCJD patients also develop cerebellar abnormalities including ataxia and, subsequently, myoclonus progressing to memory disturbances leading to severe cognitive impairment and finally akinetic mutism. Characteristic neuropathological changes include spongiform degeneration, neuronal loss, and astrogliosis in the basal ganglia and thalamus. The most striking feature is characteristic amyloid plaques of protease-resistant PrP surrounded by vacuoles in the cerebrum and cerebellum. Spread of BSE from cattle to humans is supported by evidence obtained from laboratory transmission experiments using cynomologus macaques and transgenic mice expressing human PrP^{C} [72-76]. When these animal models were inoculated with BSE and vCJD agents, strong similarities were found. Comparison revealed that patterns of infectable mouse strains, lesion profiles in the brain, proteinase-resistant PrP^{Sc} banding patterns and neuropathology were identical in animals inoculated with both agents.

Table 9.3. Cumulative number of vCJD cases reported worldwide.

Country	Total number of primary cases (number alive)[§]	Total number of secondary cases: blood transfusion (number alive)[§]
United Kingdom	165 (4)	3 (0)
France	25 (1)	-
Ireland	4 (0)	-
Italy	1 (0)	-
United States	3[†] (0)	-
Canada	1 (0)	-
Saudi Arabia	1 (1)	-
Japan	1* (0)	-
Netherlands	3 (0)	-
Portugal	2 (0)	-
Spain	5 (0)	-

[§] Number of patients diagnosed as of July 2009. [†] The third United States patient with vCJD appears to be infected as a child when living in Saudi Arabia. * The patient from Japan had resided in the United Kingdom in 1980-1996. Data retrieved from the National Creutzfeldt-Jakob Disease Surveillance Unit, United Kingdom (http://www.cjd.ed.ac.uk/).

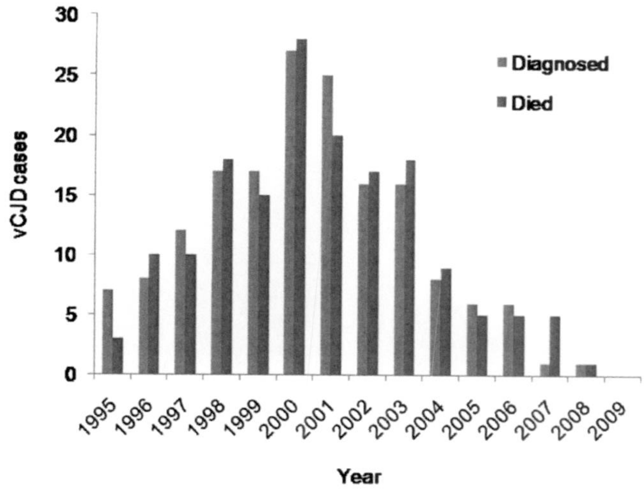

Figure 9.1. vCJD cases reported in the United Kingdom. Number of patients diagnosed (blue) and died (red) of vCJD per year. In 2009, no patient has diagnosed or died as of the end of July. The secondary transmission cases by blood transfusion are included: diagnosed in 2003 (1 case) and 2006 (2 cases); deaths in 2003, 2006, and 2007 (1 case each). Data retrieved from the National Creutzfeldt-Jakob Disease Surveillance Unit, United Kingdom (http://www.cjd.ed.ac.uk/).

The incidence of BSE and vCJD has declined and remained steadily low. The peak of vCJD incidence appears to be mid-2000. Some scientists regard the current situation as the final phase toward disappearance of vCJD [77], while others predict a multiphasic vCJD endemic with another increase of vCJD incidence affecting the human population with the heterozygous MV genotype at codon 129 [78]. At present, it is impossible to accurately predict the incidence of vCJD in the future because the incubation period of cattle to human transmission and the dose of BSE agents received by infected individuals are not known. Depending on the dose of BSE agents and susceptibility of individual human hosts, it is plausible that BSE and vCJD can transmit over a long mean incubation time in humans (over 30 years). Other concerns exist for secondary human-to-human transmission of vCJD. Some individuals exposed to BSE might be asymptomatic carriers. As exemplified in the incidence of vCJD transmission by blood transfusion [68-70], those preclinically infected humans might pose a risk of further transmission of TSEs. Because BSE agents, originally infected asymptomatic carriers, have adapted to human hosts, secondary transmission between humans might be efficient compared to the primary transmission from cattle to humans. The virulence of TSE agents might be enhanced so that incubation periods become shortened. Susceptibility relying on the *PRNP* 129 genotype might be abrogated in this type of transmission.

Animal TSes

Animal TSEs share many clinical and pathological features with human TSEs. However, the origins of agents that cause some animal TSEs such as scrapie, BSE, CWD and transmissible mink encephalopathy (TME) are difficult to elucidate. Although spontaneous emergence of TSEs, similar to that of sCJD, in certain animal species is suspected, no evidence supporting this scenario has been found. Unlike human TSEs, some animal TSEs are contagious and laterally transmissible. Feline spongiform encephalopathy (FSE) and exotic ungulate encephalopathy (EUE) are known to be occurred by secondary transmission of BSE. Animal TSEs are summarized in Table 4.

Table 9.4. Animal TSEs.

TSEs	Host	Comments
Scrapie	sheep, goat	Origin unknown, transmitted by contact with scrapie-infected animals or by possible oral exposure to tissues and environment contaminated with excreta derived from the infected animals
BSE	cattle	Origin unknown, possibly spontaneous or from scrapie, ingestion of BSE-contaminated feed
CWD	deer, elk, moose	Origin unknown, transmitted by contact with CWD-infected animals or by possible oral exposure to tissues and environment contaminated with excreta derived from the infected animals
TME	mink	Origin unknown, possibly ingestion of bovine or ovine TSE-contaminated feed, orally transmitted
FSE	cat, ocelot, Asiatic golden cat, tiger, lion, puma, cheetah	Ingestion of BSE-contaminated feed
EUE	kudu, oryx, nyala, eland, gemsbok	Ingestion of BSE-contaminated feed

BSE (bovine spongiform encephalopathy), CWD (chronic wasting disease), TME (transmissible mink encephalopathy), FSE (feline spongiform encephalopathy), EUE (exotic ungulate encephalopathy)

Scrapie

Scrapie naturally affecting sheep and goats [79] has been recognized for over two centuries. The disease develops most often in sheep with an incubation period of 1 to 2 years. Clinical signs of scrapie include ataxia, chronic pruritus, debility, recumbency, and progressive deterioration of behavior ending in death [80]. Vacuolation is a common characteristic in the brain of scrapie-infected sheep. Lesion profiling suggests that affected regions in the brain vary depending on prion strains and genotypes of sheep [81]. In all typical scrapie cases, accumulation of PrP^{Sc} is detected in the medulla oblongata, peripheral nervous system, and lymphoreticular system such as tonsils, lymph nodes, spleen, and gut-associated lymphoid tissue [82, 83]. Usually, transmission of scrapie is lateral. Although it may not be responsible for all disseminated scrapie cases, scrapie prions found in the placenta of infected ewes appear to be a source of infection for other sheep in the flock and the offspring [84]. Susceptibility to scrapie is influenced by the isolate of scrapie agent, the flock, and the breed [85, 86]. Genotyping of several flocks and breeds affected by scrapie revealed that susceptibility to scrapie is closely associated with the PrP genotype of the host [23, 87-89]. Polymorphisms in the ovine PrP gene at codons 136 (A/V), 154 (R/H), and 171 (Q/R/H) determine the susceptibility of scrapie. The VRQ allele confers the highest susceptibility, whereas the ARR allele is linked to the highest resistance. The AHQ allele is related to increased resistance but other alleles are intermediate. Although the mechanism underlying the susceptibility of sheep with these allelic variants is not certain, these effects appear to depend on the conversion efficiencies of the respective PrP^C to PrP^{Sc}.

In addition, a novel scrapie type, designated Nor98, was reported in Norway in 1998 and subsequently found in other countries [90]. This unusual scrapie presents with ataxia but no obvious pruritus. Atypical Nor 98 scrapie also differs from classic scrapie in the distribution of pathological changes and PrP^{Sc} in the CNS, and the Western blot profile of PrP^{Sc}. Furthermore, the PrP genotypes affected most frequently and the age of the animals at the time of detection were different in Nor98. Unlike typical scrapie, PrP^{Sc} preferentially accumulated in the cerebellar cortex and cerebrum, but not in the lymphoreticular tissues. Vacuolation also affected the cerebellar and cerebral cortices, but was less prominent in the medulla oblongata. Moreover, Nor98 showed an unusual ratio in the glycosylation pattern of PrP^{Sc} and a small size PrP^{Sc} band in Western blot analysis. The affected animals were older with the mean age at onset over 7 to 9 years than those affected by typical scrapie most often occurring at 2 to 5 years of age. Nor98 affected sheep with the AHQ alleles that are more resistant and less associated with classic scrapie. Moreover, the $AF_{141}RQ$ allele conferred a higher risk than the AHQ or $AL_{141}RQ$ alleles [91].

BSE

BSE is a fatal, progressive neurodegenerative disorder affecting cattle at an average age of 4 to 5 years [92]. Since the first appearance of BSE in the mid-1980s, nearly 200,000 cases have been identified in the United Kingdom with the peak of the epidemic occurring between 1992 and1993 (Table 5 and Figure 9.2). The geographic distribution of BSE has increased to 25 countries, including the United States and Canada (Table 5).

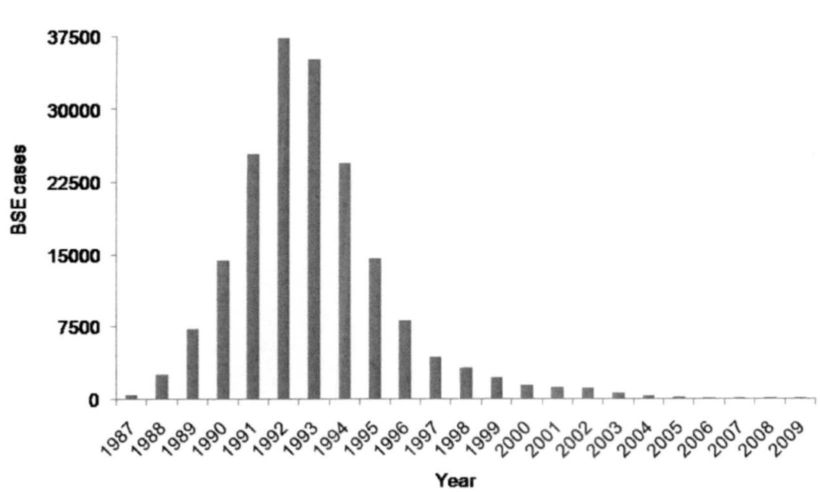

Figure 9.2. BSE cases reported in the United Kingdom. In 2009, the BSE cases were counted until July. Data were retrieved from the World Organization for Animal Health (Office International des Epizooties, OIE; http://www.oie.int/eng/info/en_esb.htm).

The primary clinical signs of disease include hyperactivity, apprehension, gait disturbance, behavioral abnormalities, weakness, and recumbency [93]. BSE presents as highly concentrated lesions in the brain stem [94]. Bilaterally symmetric vacuolar lesions are found in the gray matter neuropil. However, few amyloid plaques are found in damaged tissues. The rare plaques are found mostly in the thalamus. Distribution of PrP^{Sc} accumulation in the brain coincides with the area of spongiosis [94]. In particular, a limited accumulation of PrP^{Sc} is also detected in the Peyer's patches in the distal portion of the ileum [95].

Neither lateral nor vertical transmission is evident. The BSE epidemic was caused by feeding of meat and bone meal contaminated with the rendered tissues of BSE-positive cattle, which allowed the infectious agent to spread. The incidence of BSE has declined dramatically since the ban of meat and bone meal from feeding. However, BSE has crossed the species barrier and transmitted to other species. vCJD in humans has been transmitted by ingesting a BSE-contaminated diet, and has killed 165 individuals in the United Kingdom and 46 elsewhere (Table 3). It is estimated that 460,000 to 482,000 BSE-infected animals had entered the human food chain before controls on high-risk offal [96]. With ingestion of contaminated meat and bone meal, natural secondary BSE infection has also occurred in several species of ruminants (including exotic ungulates), cats, and nonhuman primates in zoos [97-99], as well as in domestic cats [100]. The similarity in pathology and in laboratory tests, including PrP^{Sc} banding patterns in Western blots and comparative titration in various mouse strains, supports transmission of BSE to humans and other affected species. However, the origin of BSE agents is unknown. It remains to be answered whether BSE originated by adaptation from an unusual strain of sheep scrapie or from a rare, unrecognized idiopathic bovine TSE case.

Table 9.5. Cumulative BSE cases reported worldwide.

Country	Total number of cases	Country	Total number of cases
Austria	8	Liechtenstein	16
Belgium	265	Luxembourg	521
Canada	34	Netherlands	68
Czech Republic	56	Poland	112
Denmark	26	Portugal	52
Finland	2	Slovakia	4
France	1958	Slovenia	3867
Germany	760	Spain	1520
Greece	2	Sweden	4
Ireland	2901	Switzerland	5628
Israel	2	United Kingdom	184594
Italy	280	United States	560
Japan	70		

The numbers represent the cumulative cases of BSE in each country as of July 2009. Data were retrieved from the World Organization for Animal Health (Office International des Epizooties, OIE; http://www.oie.int/eng/info/en_esb.htm).

Unlike scrapie, it has been hypothesized that a single strain exists for BSE agents. Recently, multiple cases of atypical BSE were identified in Italy and France [101, 102]. The unusual features of novel BSE, designated bovine amyloidotic spongiform encephalopathy (BASE), include the presence of amyloid plaques predominantly in the white matter, and reduced PrP^{Sc} accumulation in the brain stem. Also, the atypical BASE cases showed a different distribution of brain lesions and a different molecular signature of proteinase-resistant PrP^{Sc} compared to typical BSE cases. Since the molecular mass of the unglycosylated PrP^{Sc} band was lower compared to typical BSE, these atypical BSE cases are categorized as L-types. Interestingly, the L-types resembled sCJD in the electrophoretic profile of PrP^{Sc} [102]. Another group of atypical BSEs, termed H-type [103, 104], presents with a higher molecular mass of the unglycosylated PrP^{Sc} band than typical BSE. Like L-type, H-type is transmissible to laboratory animals and maintains its own characteristic H-type molecular signatures. However, it shows different neuropathology from typical BSE and L-type. Additionally, atypical biochemical characteristics of PrP^{Sc} were also shown in a Japanese case [105]. Furthermore, passages of BSE PrP^{Sc} to mice suggested that two distinct BSE strains may exist in British cattle [106]. These studies proposed that BSE exists in multiple strains and that a spontaneous or sporadic form of TSE may occur in cattle as shown in sCJD in humans and Nor98 in sheep.

CWD

CWD is a TSE affecting mule deer, white tail deer, Rocky Mountain elk, moose and possibly black tail deer [107]. Currently, CWD is a serious problem in captive and wild populations of deer and elk in North America. In 1967, this disease was first recognized in captive deer at a research facility in Colorado [108]. By mid-1980s, CWD had been identified in free-ranging cervids in Colorado and Wyoming. Currently, CWD has spread to both captive and wild cervid populations in 15 states in the United States and two Canadian provinces [109]. In certain wild populations of deer in northern Colorado and southern Wyoming, the incidence of CWD is more than 10%. Additionally, CWD outbreaks have been reported in farmed elk imported into South Korea from Canada [110]. Clinically, CWD manifests behavioral changes (depression, somnolence, and aggression); progressive weight loss; excessive salivation; polydipsia; polyuria; tremors; ataxia; paralysis; weakness; inability to stand; inability to swallow; dehydration; dull hair coat; and emaciation [108, 111]. The age of clinical onset in elk ranges from 2 to 8 years, and the duration of symptoms ranges from 5 to 12 months before death [112]. Pathology of CWD is characterized by neuronal degeneration, spongiform vacuolation, astrocytic hypertrophy, and hyperplasia. Amyloid plaques are common in deer brains [113]. Microscopic lesions and neuropathology of CWD are concentrated to the medulla oblongata, olfactory bulb, cortex, and hypothalamus. Accumulation of PrP^{Sc} has been observed in the brain, tonsils, lymph nodes, the small and large intestine, Peyer's patches, and spleen of CWD-affected deer [114]. PrP^{Sc} accumulates in the lymphoid tissue before it is detected in the brain stem [107]. Polymorphisms of the *PRNP* gene at codons 95 (Q/H), 96 (G/S), 116 (A/G), 132 (M/L), 138 (S/N) and 225 (S/F) are partly associated with susceptibility to CWD [115]. The 96SS, 132 LL, and 225FF genotypes was linked to a lower risk for CWD infection, whereas animals with 96GG, 132MM, and 225SS were more susceptible to CWD. The origin of CWD is unknown. However, it has been postulated that CWD is the result of a spontaneous naturally occurring TSE of cervids or originated by the transmission of scrapie to cervids. In fact, inoculation of elk with scrapie resulted in CWD-like conditions [116]. CWD is laterally transmitted by both direct and indirect contact [117]. Pastures contaminated by excreta and saliva are a likely cause to result in transmission and spread of CWD [112, 118].

Other Animal TSEs

Other known animal TSEs, such as EUE and FSE, occurred as a result of BSE transmission to respective species. As described, EUE in ungulates and FSE in domestic cats and other feline species in zoos were caused by feeding BSE-contaminated material. Similarly, non-human primates in a zoo and a research facility developed TSEs by ingesting diets that included meat and bone meal supplements contaminated with BSE agents. TME outbreaks have been reported in several mink ranches in the United States and a few other countries [119]. However, the origin of TME largely remains uncertain. Animal tissues from either scrapie-infected sheep or TSE-infected cattle were believed to be a plausible source for the infection. In the most recent TME outbreak in the United States, downer cattle used as the major source of mink food was speculated to provide TSE agents [119]. In this case, cattle TSE agents transmitted to mink appear to be different from those found in BSE from Europe.

Conclusion

Due to the peculiar infectious nature of pathogenic agents and striking transmission of BSE to humans, TSEs have attracted enormous attention from diverse areas of our society. In particular, research in the past decades has revealed the unique properties of prions and the extraordinary biology involved in them. We now understand the species barrier and prion strains involved in transmission of TSEs. Accumulating data have provided information for the understanding of the factors influencing prion pathogenesis. Identifying the etiology, clinical and pathological features, and the mode of transmission of individual TSEs in several species has broadened our knowledge to answer the mystery of transmission and emergence of new diseases. Nevertheless, detailed mechanisms that facilitate prion replication and pathogenesis are unknown. Furthermore, secondary human-to-human transmission of vCJD through unexpected routes, emergence of atypical BSE and scrapie, and potential risks of CWD to humans propose more questions to be answered. Although the outbreaks of BSE and vCJD seem to be controlled, the potential risk of TSEs to human health still remains. Interdisciplinary efforts to investigate molecular aspects of prion conversion, replication, pathogenesis, and transmission are needed.

Acknowledgments

I am grateful to Paula Thomason for editorial assistance. This work was supported in part by funds from the Sanders-Brown Center on Aging and College of Medicine, University of Kentucky.

References

[1] Prusiner SB. Prions. *Proc Natl Acad Sci USA*, 1998, 95, 13363-13383.

[2] Anonymous. Bovine spongiform encephalopathy. Paris, France OIE-World Organisation for Animal Health; [cited 2009 August 15]; Available from: *http://www.oie.int/eng/info/en_esb.htm.*

[3] Prusiner SB. Novel proteinaceous infectious particles cause scrapie. *Science*, 1982, 216, 136-144.

[4] Bruce ME. TSE strain variation: An investigation into prion disease diversity. *Br Med Bull*, 2003, 66, 99-108.

[5] Scott M; Peretz D; Ridley RM; Baker HF; DeArmond SJ, Prusiner SB. *Transgenetic investigations of the species barrier and prion strains*. In: Prusiner SB, editor. Prion biology and diseases. 2nd ed. Cold Spring Harbor: Cold Spring Harbor Laboratory Press; 2004. p. 435-482.

[6] Scott M; Foster D; Mirenda C; Serban D; Coufal F; Walchli M; Torchia M; Groth D; Carlson G; DeArmond SJ; Westaway D, Prusiner SB. Transgenic mice expressing hamster prion protein produce species-specific scrapie infectivity and amyloid plaques. *Cell*, 1989, 59, 847-857.

[7] Kimberlin RH; Cole S, Walker CA. Temporary and permanent modifications to a single strain of mouse scrapie on transmission to rats and hamsters. *J Gen Virol*, 1987, 68, 1875-1881.

[8] Fraser H, Dickinson AG. The sequential development of the brain lesions of scrapie in three strains of mice. *J Comp Pathol*, 1968, 78, 301-311.

[9] Will RG; Ironside JW; Zeidler M; Cousens SN; Estibeiro K; Alperovitch A; Poser S; Pocchiari M; Hofman A, Smith PG. A new variant of Creutzfeldt-Jakob disease in the UK. *Lancet*, 1996, 347, 921-925.

[10] Dickinson AG; Meikle VMH, Fraser H. Identification of a gene which controls the incubation period of some strains of scrapie agent in mice. *J Comp Pathol*, 1968, 78, 293-299.

[11] Carlson GA; Kingsbury DT; Goodman PA; Coleman S; Marshall ST; DeArmond S; Westaway D, Prusiner SB. Linkage of prion protein and scrapie incubation time genes. *Cell*, 1986, 46, 503-511.

[12] Hunter N; Hope J; McConnell I, Dickinson AG. Linkage of the scrapie-associated fibril protein (PrP) gene and Sinc using congenic mice and restriction fragment length polymorphism analysis. *J Gen Virol*, 1987, 68, 2711-2716.

[13] Bueler H; Aguzzi A; Sailer A; Greiner R-A; Autenried P; Aguet M, Weissmann C. Mice devoid of PrP are resistant to scrapie. *Cell*, 1993, 73, 1339-1347.

[14] Manson JC; Clarke AR; McBride PA; McConnell I, Hope J. PrP gene dosage determines the timing but not the final intensity or distribution of lesions in scrapie pathology. *Neurodegeneration*, 1994, 3, 331-340.

[15] Prusiner SB; Groth D; Serban A; Koehler R; Foster D; Torchia M; Burton D; Yang S-L, DeArmond SJ. Ablation of the prion protein (PrP) gene in mice prevents scrapie and facilitates production of anti-PrP antibodies. *Proc Natl Acad Sci USA*, 1993, 90, 10608-10612.

[16] Telling GC; Haga T; Torchia M; Tremblay P; DeArmond SJ, Prusiner SB. Interactions between wild-type and mutant prion proteins modulate neurodegeneration in transgenic mice. *Genes Dev*, 1996, 10, 1736-1750.

[17] Fischer M; Rulicke T; Raeber A; Sailer A; Moser M; Oesch B; Brandner S; Aguzzi A, Weissmann C. Prion protein (PrP) with amino-proximal deletions restoring susceptibility of PrP knockout mice to scrapie. *EMBO J*, 1996, 15, 1255-1264.

[18] Kimberlin RH, Walker CA. Evidence that the transmission of one source of scrapie agent to hamsters involves separation of agent strains from a mixture. *J Gen Virol*, 1978, 39, 487-496.

[19] Green KM; Castilla J; Seward TS; Napier DL; Jewell JE; Soto C, Telling GC. Accelerated high fidelity prion amplification within and across prion species barriers. *PLoS Pathog*, 2008, 4, e1000139.

[20] Browning SR; Mason GL; Seward T; Green M; Eliason GA; Mathiason C; Miller MW; Williams ES; Hoover E, Telling GC. Transmission of prions from mule deer and elk with chronic wasting disease to transgenic mice expressing cervid PrP. *J Virol*, 2004, 78, 13345-13350.

[21] Bossers A; Belt PBGM; Raymond GJ; Caughey B; de Vries R, Smits MA. Scrapie susceptibility-linked polymorphisms modulate the *in vitro* conversion of sheep prion protein to protease-resistant forms. *Proc Natl Acad Sci USA*, 1997, 94, 4931-4936.

[22] Bossers A; Schreuder BE; Muileman IH; Belt PB, Smits MA. PrP genotype contributes to determining survival times of sheep with natural scrapie. *J Gen Virol*, 1996, 77, 2669-2673.

[23] Goldmann W; Hunter N; Foster JD; Salbaum JM; Beyreuther K, Hope J. Two alleles of a neural protein gene linked to scrapie in sheep. *Proc Natl Acad Sci USA*, 1990, 87, 2476-2480.

[24] Belt PB; Muileman IH; Schreuder BEC; Ruijter JB; Gielkens ALJ, Smits MA. Identification of five allelic variants of the sheep PrP gene and their association with natural scrapie. *J Gen Virol*, 1995, 76, 509-517.

[25] Elsen J-M; Amigues Y; Schelcher F; Ducrocq V; Andreoletti O; Eychenne F; Tien Khang JV; Poivey J-P; Lantier F, Laplanche J-L. Genetic susceptibility and transmission factors in scrapie: detailed analysis of an epidemic in a closed flock of Romanov. *Arch Virol*, 1999, 144, 431-445.

[26] Telling GC; Parchi P; DeArmond SJ; Cortelli P; Montagna P; Gabizon R; Mastrianni J; Lugaresi E; Gambetti P, Prusiner SB. Evidence for the conformation of the pathologic isoform of the prion protein enciphering and propagating prion diversity. *Science*, 1996, 274, 2079-2082.

[27] Peretz D; Williamson RA; Legname G; Matsunaga Y; Vergara J; Burton D; DeArmond SJ; Prusiner SB, Scott MR. A change in the conformation of prions accompanies the emergence of a new prion strain. *Neuron*, 2002, 34, 921-932.

[28] Safar J; Wille H; Itri V; Groth D; Serban H; Torchia M; Cohen FE, Prusiner SB. Eight prion strains have PrP^{Sc} molecules with different conformations. *Nat Med*, 1998, 4, 1157-1165.

[29] Collinge J; Sidle KCL; Meads J; Ironside J, Hill AF. Molecular analysis of prion strain variation and the aetiology of "new variant" CJD. *Nature*, 1996, 383, 685-690.

[30] Parchi P; Castellani R; Capellari S; Ghetti B; Young K; Chen SG; Farlow M; Dickson DW; Sima AAF; Trojanowski JQ; Petersen RB, Gambetti P. Molecular basis of

phenotypic variability in sporadic Creutzfeldt-Jakob disease. *Ann Neurol*, 1996, 39, 767-778.

[31] Ryou C. Prions and prion diseases: Fundamentals and mechanistic details. *J Microbiol Biotechnol* 2007, 17, 1059-1070.

[32] McKinley MP; Meyer RK; Kenaga L; Rahbar F; Cotter R; Serban A, Prusiner SB. Scrapie prion rod formation *in vitro* requires both detergent extraction and limited proteolysis. *J Virol*, 1991, 65, 1340-1351.

[33] Prusiner SB; McKinley MP; Bowman KA; Bolton DC; Bendheim PE; Groth DF, Glenner GG. Scrapie prions aggregate to form amyloid-like birefringent rods. *Cell*, 1983, 35, 349-358.

[34] Merz PA; Somerville RA; Wisniewski HM, Iqbal K. Abnormal fibrils from scrapie-infected brain. *Acta Neuropathol (Berl)*, 1981, 54, 63-74.

[35] Bucciantini M; Giannoni E; Chiti F; Baroni F; Formigli L; Zurdo J; Taddei N; Ramponi G; Dobson CM, Stefani M. Inherent toxicity of aggregates implies a common mechanism for protein misfolding diseases. *Nature*, 2002, 416, 507-511.

[36] Caughey B, Lansbury Jr PT. Protofibirils, pores, fibrils, and neurodegeneration: Separating the responsible protein aggregates from the innocent bystanders. *Annu Rev Neurosci*, 2003, 26, 267-298.

[37] DeArmond SJ; Mobley WC; DeMott DL; Barry RA; Beckstead JH, Prusiner SB. Changes in the localization of brain prion proteins during scrapie infection. *Neurology*, 1987, 37, 1271-1280.

[38] Hetz C; Russelakis-Carneiro M; Maundrell K; Castilla J, Soto C. Caspase-12 and endoplasmic reticulum stress mediate neurotoxicity of pathological prion protein. *EMBO J*, 2003, 22, 5435-5445.

[39] Marella M; Gaggioli C; Batoz M; Deckert M; Tartare-Deckert S, Chabry J. Pathological prion protein exposure switches on neuronal mitogen-activated protein kinase pathway resulting in microglia recruitment. *J Biol Chem*, 2005, 280, 1529-1534.

[40] Kazlauskaite J; Young A; Gardner CE; Macpherson JV; Vénien-Bryan C, Pinheiro TJT. An unusual soluble [beta]-turn-rich conformation of prion is involved in fibril formation and toxic to neuronal cells. *Biochem Biophys Res Commun*, 2005, 328, 292-305.

[41] Novitskaya V; Bocharova OV; Bronstein I, Baskakov IV. Amyloid fibrils of mammalian prion protein are highly toxic to cultured cells and primary neurons. *J Biol Chem*, 2006, 281, 13828-13836.

[42] Brandner S; Isenmann S; Raeber A; Fischer M; Sailer A; Kobayashi Y; Marino S; Weissmann C, Aguzzi A. Normal host prion protein necessary for scrapie-induced neurotoxicity. *Nature*, 1996, 379, 339-343.

[43] Kuwahara C; Takeuchi AM; Nishimura T; Haraguchi K; Kubosaki A; Matsumoto Y; Saeki K; Matsumoto Y; Yokoyama T; Itohara S, Onodera T. Prions prevent neuronal cell-line death. *Nature*, 1999, 400, 225-226.

[44] Aguzzi A, Heikenwalder M. Prions, cytokines, and chemokines: A meeting in lymphoid organs. *Immunity*, 2005, 22, 145-154.

[45] Prinz M; Huber G; Macpherson AJS; Heppner FL; Glatzel M; Eugster H-P; Wagner N, Aguzzi A. Oral prion infection requires normal numbers of Peyer's patches but not of enteric lymphocytes. *Am J Pathol*, 2003, 162, 1103-1111.

[46] Kimberlin RH, Walker CA. Pathogenesis of mouse scrapie: dynamics of agent replication in spleen, spinal cord and brain after infection by different routes. *J Comp Pathol*, 1979, 89, 551-562.
[47] Fraser H, Dickinson AG. Pathogenesis of scrapie in the mouse: the role of the spleen. *Nature*, 1970, 226, 462-463.
[48] Race R; Oldstone M, Chesebro B. Entry versus blockade of brain infection following oral or intraperitoneal scrapie administration: role of prion protein expression in peripheral nerves and spleen. *J Virol*, 2000, 74, 828-833.
[49] Klein MA; Frigg R; Flechsig E; Raeber AJ; Kalinke U; Bluethmann H; Bootz F; Suter M; Zinkernagel RM, Aguzzi A. A crucial role for B cells in neuroinvasive scrapie. *Nature*, 1997, 390, 687-691.
[50] Montrasio F; Frigg R; Glatzel M; Klein MA; Mackay F; Aguzzi A, Weissmann C. Impaired prion replication in spleens of mice lacking functional follicular dendritic cells. *Science*, 2000, 288, 1257-1259.
[51] Mabbott NA; Bruce ME; Botto M; Walport MJ, Pepys MB. Temporary depletion of complement component C3 or genetic deficiency of C1q significantly delays onset of scrapie. *Nat Med*, 2001, 7, 485-487.
[52] Mabbott NA; Young J; McConnell I, Bruce ME. Follicular dendritic cell dedifferentiation by treatment with an inhibitor of the lymphotoxin pathway dramatically reduces scrapie susceptibility. *J Virol*, 2003, 77, 6845-6854.
[53] Alpers MP. The epidemiology of kuru: monitoring the epidemic from its peak to its end. *Philos Trans R Soc Lond B Biol Sci*, 2008, 363, 3707-3713.
[54] Prusiner SB. Shattuck Lecture: Neurodegenerative diseases and prions. *N Engl J Med*, 2001, 344, 1516-1526.
[55] DeArmond SJ; Ironside JW; Bouzamondo-Bernstein E; Peretz D, Fraser JR. *Neuropathology of prion diseases*. In: Prusiner SB, editor. Prion biology and diseases. 2nd ed. Cold Spring Harbor: Cold Spring Harbor Laboratory Press; 2004. p. 777-856.
[56] Creutzfeldt HG. Uber eine eigenartige herdformige Erkrankung des Zentralnervensystems. *Z Gesamte Neurol Psychiatrie*, 1920, 57, 1-18.
[57] Jakob A. Uber eigenartige Erkrankungen des Zentralnervensystems mit bemerkenswertem anatomischen Befunde (spastische Pseudosklerose-Encephalo-myelopathie mit disseminierten Degenerationsherden). *Z Gesamte Neurol Psychiatrie*, 1921, 64, 147-228.
[58] Gerstmann J; Sträussler E, Scheinker I. Uber eine eigenartige hereditar- familiare Erkrankung des Zentralnervensystems zugleich ein Beitrag zur frage des vorzeitigen lokalen Alterns. *Z Neurol*, 1936, 154, 736-762.
[59] Gajdusek DC, Zigas V. Degenerative disease of the central nervous system in New Guinea: The endemic occurrence of "kuru" in the native population. *N Engl J Med*, 1957, 257, 974-978.
[60] Gambetti P; Kong Q; Zou W; Parchi P, Chen SG. Sporadic and familial CJD: classification and characterisation. *Br Med Bull*, 2003, 66, 213-239.
[61] Montagna P; Gambetti P; Cortelli P, Lugaresi E. Familial and sporadic fatal insomnia. *Lancet Neurol*, 2003, 2, 167-176.
[62] Kong Q; Surewicz WK; Petersen RB; Zou W; Chen SG; Gambetti P; Parchi P; Capellari S; Goldfarb L; Montagna P; Lugaresi E; Piccardo P, Ghetti B. *Inherited prion*

diseases. In: Prusiner SB, editor. Prion biology and diseases. 2nd ed. Cold Spring Harbor: Cold Spring Harbor Laboratory Press; 2004. p. 673-775.

[63] Duffy P; Wolf J; Collins G; DeVoe AG; Streeten B, Cowen D. Possible person-to-person transmission of Creutzfeldt-Jakob disease. *N Engl J Med*, 1974, 290, 692-693.

[64] Brown P; Preece M; Brandel JP; Sato T; McShane L; Zerr I; Fletcher A; Will RG; Pocchiari M; Cashman NR; d'Aignaux JH; Cervenakova L; Fradkin J; Schonberger LB, Collins SJ. Iatrogenic Creutzfeldt-Jakob disease at the millennium. *Neurology*, 2000, 55, 1075-1081.

[65] Thadani V; Penar PL; Partington J; Kalb R; Janssen R; Schonberger LB; Rabkin CS, Prichard JW. Creutzfeldt-Jakob disease probably acquired from a cadaveric dura mater graft. Case report. *J Neurosurg*, 1988, 69, 766-769.

[66] Koch TK; Berg BO; DeArmond SJ, Gravina RF. Creutzfeldt-Jakob disease in a young adult with idiopathic hypopituitarism: Possible relation to the administration of cadaveric human growth hormone. *N Engl J Med*, 1985, 313, 731-733.

[67] Bernoulli C; Siegfried J; Baumgartner G; Regli F; Rabinowicz T; Gajdusek DC, Gibbs CJ, Jr. Danger of accidental person-to-person transmission of Creutzfeldt-Jakob disease by surgery. *Lancet*, 1977, 1, 478-479.

[68] Llewelyn CA; Hewitt PE; Knight RS; Amar K; Cousens S; Mackenzie J, Will RG. Possible transmission of variant Creutzfeldt-Jakob disease by blood transfusion. *Lancet*, 2004, 363, 417-421.

[69] Peden AH; Head MW; Ritchie DL; Bell JE, Ironside JW. Preclinical vCJD after blood transfusion in a *PRNP* codon 129 heterozygous patient. *Lancet*, 2004, 364, 527-529.

[70] Wroe SJ; Pal S; Siddique D; Hyare H; Macfarlane R; Joiner S; Linehan JM; Brandner S; Wadsworth JDF; Hewitt P, Collinge J. Clinical presentation and pre-mortem diagnosis of variant Creutzfeldt-Jakob disease associated with blood transfusion: a case report. *Lancet*, 2006, 368, 2061-2067.

[71] Collinge J; Beck J; Campbell T; Estibeiro K, Will RG. Prion protein gene analysis in new variant cases of Creutzfeldt-Jakob disease. *Lancet*, 1996, 348, 56.

[72] Bruce ME; Will RG; Ironside JW; McConnell I; Drummond D; Suttie A; McCardle L; Chree A; Hope J; Birkett C; Cousens S; Fraser H, Bostock CJ. Transmissions to mice indicate that 'new variant' CJD is caused by the BSE agent. *Nature*, 1997, 389, 498-501.

[73] Hill AF; Desbruslais M; Joiner S; Sidle KCL; Gowland I; Collinge J; Doey LJ, Lantos P. The same prion strain causes vCJD and BSE. *Nature*, 1997, 389, 448-450.

[74] Lasmezas CI; Deslys J-P; Demaimay R; Adjou KT; Lamoury F; Dormont D; Robain O; Ironside J, Hauw J-J. BSE transmission to macaques. *Nature*, 1996, 381, 743-744.

[75] Lasmezas CI; Fournier J-G; Nouvel V; Boe H; Marce D; Lamoury F; Kopp N; Hauw J-J; Ironside J; Bruce M; Dormont D, Deslys J-P. Adaptation of the bovine spongiform encephalopathy agent to primates and comparison with Creutzfeldt- Jakob disease: Implications for human health. *Proc Natl Acad Sci USA*, 2001, 98, 4142-4147.

[76] Scott MR; Will R; Ironside J; Nguyen H-OB; Tremblay P; DeArmond SJ, Prusiner SB. Compelling transgenetic evidence for transmission of bovine spongiform encephalopathy prions to humans. *Proc Natl Acad Sci USA*, 1999, 96, 15137-15142.

[77] Andrews NJ; Farrington CP; Ward HJT; Cousens SN; Smith PG; Molesworth AM; Knight RSG; Ironside JW, Will RG. Deaths from variant Creutzfeldt-Jakob disease in the UK. *Lancet*, 2003, 361, 751-752.

[78] Collinge J; Whitfield J; McKintosh E; Beck J; Mead S; Thomas DJ, Alpers MP. Kuru in the 21st century--an acquired human prion disease with very long incubation periods. *Lancet*, 2006, 367, 2068-2074.

[79] Wood JL; Lund LJ, Done SH. The natural occurrence of scrapie in moufflon. *Vet Rec*, 1992, 130, 25-27.

[80] Parry HB. Scrapie disease in sheep. Oppenheimer DR, editor. New York: Academic Press; 1983.

[81] Ligios C; Jeffrey M; Ryder SJ; Bellworthy SJ, Simmons MM. Distinction of scrapie phenotypes in sheep by lesion profiling. *J Comp Pathol*, 2002, 127, 45-57.

[82] Miller JM; Jenny AL; Taylor WD; Marsh RF; Rubenstein R, Race RE. Immunohistochemical detection of prion protein in sheep with scrapie. *J Vet Diagn Invest*, 1993, 5, 309-316.

[83] van Keulen LJ; Schreuder BE; Meloen RH; Mooij-Harkes G; Vromans ME, Langeveld JP. Immunohistochemical detection of prion protein in lymphoid tissues of sheep with natural scrapie. *J Clin Microbiol*, 1996, 34, 1228-1231.

[84] Caplazi P; O'Rourke K; Wolf C; Shaw D, Baszler T. Biology of PrP^{Sc} accumulation in two natural scrapie-infected sheep flocks. *J Vet Diagn Invest*, 2004, 16, 489-496.

[85] Dickinson AG. *Scrapie in sheep and goats*. In: Kimberlin RH, editor. Slow virus diseases of animals and man. Amsterdam: North-Holland Publishing; 1976. p. 209-241.

[86] Pattison IH. The relative susceptibility of sheep, goats and mice to two types of the goat scrapie agent. *Res Vet Sci*, 1966, 7, 207-212.

[87] Goldmann W; Hunter N; Benson G; Foster JD, Hope J. Different scrapie-associated fibril proteins (PrP) are encoded by lines of sheep selected for different alleles of the *Sip* gene. *J Gen Virol*, 1991, 72, 2411-2417.

[88] Westaway D; Zuliani V; Cooper CM; Da Costa M; Neuman S; Jenny AL; Detwiler L, Prusiner SB. Homozygosity for prion protein alleles encoding glutamine-171 renders sheep susceptible to natural scrapie. *Genes Dev*, 1994, 8, 959-969.

[89] Hunter N; Goldmann W; Smith G, Hope J. The association of a codon 136 PrP gene variant with the occurrence of natural scrapie. *Arch Virol*, 1994, 137, 171-177.

[90] Benestad SL; Sarradin P; Thu B; Schönheit J; Tranulis MA; Bratberg B. Cases of scrapie with unusual features in Norway and designation of a new type, Nor98. *Vet Rec*, 2003, 153, 202-208.

[91] Moum T; Olsaker I; Hopp P; Moldal T; Valheim M; Moum T, Benestad SL. Polymorphisms at codons 141 and 154 in the ovine prion protein gene are associated with scrapie Nor98 cases. *J Gen Virol*, 2005, 86, 231-235.

[92] Wells GAH; Scott AC; Johnson CT; Gunning RF; Hancock RD; Jeffrey M; Dawson M, Bradley R. A novel progressive spongiform encephalopathy in cattle. *Vet Rec*, 1987, 121, 419-420.

[93] Konold T; Bone G; Ryder S; Hawkins SAC; Courtin F, Berthelin-Baker C. Clinical findings in 78 suspected cases of bovine spongiform encephalopathy in Great Britain. *Vet Rec*, 2004, 155, 659-666.

[94] Wells GAH, Wilesmith JW. The neuropathology and epidemiology of bovine spongiform encephalopathy. *Brain Pathol*, 1995, 5, 91-103.

[95] Terry LA; Marsh S; Ryder SJ; Hawkins SA; Wells GA, Spencer YI. Detection of disease-specific PrP in the distal ileum of cattle exposed orally to the agent of bovine spongiform encephalopathy. *Vet Rec*, 2003, 152, 387-392.

[96] Valleron A-J; Boelle P-Y; Will R, Cesbron J-Y. Estimation of epidemic size and incubation time based on age characteristics of vCJD in the United Kingdom. *Science*, 2001, 294, 1726-1728.

[97] Kirkwood JK, Cunningham AA. Epidemiological observations on spongiform encephalopathies in captive wild animals in the British Isles. *Vet Rec*, 1994, 135, 296-303.

[98] Willoughby K; Kelly DF; Lyon DG, Wells GAH. Spongiform encephalopathy in a captive puma (*Felis concolor*). *Vet Rec*, 1992, 131, 431-434.

[99] Bons N; Mestre-Frances N; Belli P; Cathala F; Gajdusek DC, Brown P. Natural and experimental oral infection of nonhuman primates by bovine spongiform encephalopathy agents. *Proc Natl Acad Sci USA*, 1999, 96, 4046-4051.

[100] Wyatt JM; Pearson GR; Smerdon TN; Gruffydd-Jones TJ; Wells GAH, Wilesmith JW. Naturally occurring scrapie-like spongiform encephalopathy in five domestic cats. *Vet Rec*, 1991, 129, 233-236.

[101] Biacabe AG; Laplanche JL; Ryder S, Baron T. Distinct molecular phenotypes in bovine prion diseases. *EMBO Rep*, 2004, 5, 110-115.

[102] Casalone C; Zanusso G; Acutis P; Ferrari S; Capucci L; Tagliavini F; Monaco S, Caramelli M. Identification of a second bovine amyloidotic spongiform encephalopathy: Molecular similarities with sporadic Creutzfeldt-Jakob disease. *Proc Natl Acad Sci USA*, 2004, 101, 3065-3070.

[103] Biacabe A-G; Morignat E; Vulin J; Calavas D, Baron TGM. Atypical bovine spongiform encephalopathies, France, 2001-2007. *Emerg Infect Dis,* 2008, 14, 298-300.

[104] Terry LA; Jenkins R; Thorne L; Everest SJ; Chaplin MJ; Davis LA, Stack MJ. First case of H-type bovine spongiform encephalopathy identified in Great Britain. *Vet Rec*, 2007, 160, 873-874.

[105] Yamakawa Y; Hagiwara K; Nohtomi K; Nakamura Y; Nishijima M; Higuchi Y; Sato Y, Sata T. Atypical proteinase K-resistant prion protein (PrP^{res}) observed in an apparently healthy 23-month-old Holstein steer. *Jpn J Infect Dis*, 2003, 56, 221-222.

[106] Lloyd SE; Linehan JM; Desbruslais M; Joiner S; Buckell J; Brandner S; Wadsworth JDF, Collinge J. Characterization of two distinct prion strains derived from bovine spongiform encephalopathy transmissions to inbred mice. *J Gen Virol*, 2004, 85, 2471-2478.

[107] Williams ES, Miller MW. Chronic wasting disease in deer and elk in North America. *Rev Sci Tech*, 2002, 21, 305-316.

[108] Williams ES, Young S. Chronic wasting disease of captive mule deer: a spongiform encephalopathy. *J Wildl Dis*, 1980, 16, 89-98.

[109] Anonymous. Chronic wasting disease. Madison, Wisconsin: National Wildlife Health Center, United State Geological Survey; [cited 2009 Augsut 15]; Available from: http://www.nwhc.usgs.gov/disease_information/chronic_wasting_disease/index.jsp.

[110] Sohn HJ; Kim JH; Choi KS; Nah JJ; Joo YS; Jean YH; Ahn SW; Kim OK; Kim DY, Balachandran A. A case of chronic wasting disease in an elk imported to Korea from Canada. *J Vet Med Sci*, 2002, 64, 855-858.

[111] Spraker TR; Miller MW; Williams ES; Getzy DM; Adrian WJ; Schoonveld GG; Spowart RA; O'Rourke KI; Miller JM, Merz PA. Spongiform encephalopathy in free-ranging mule deer (*Odocoileus hemionus*), white-tailed deer (*Odocoileus virginianus*),

and Rocky Mountain elk (*Cervus elaphus nelsoni*) in northcentral Colorado. *J Wildl Dis*, 1997, 33, 1-6.

[112] Miller MW; Wild MA, Williams ES. Epidemiology of chronic wasting disease in captive Rocky Mountain elk. *J Wildl Dis*, 1998, 34, 532-538.

[113] Williams ES, Young S. Neuropathology of chronic wasting disease of mule deer (*Odocoileus hemionus*) and Elk (Cervus elaphus nelsoni). *Vet Pathol*, 1993, 30, 36-45.

[114] Sigurdson CJ; Williams ES; Miller MW; Spraker TR; O'Rourke KI, Hoover EA. Oral transmission and early lymphoid tropism of chronic wasting disease PrP^{res} in mule deer fawns (*Odocoileus hemionus*). *J Gen Virol*, 1999, 80, 2757-2764.

[115] O'Rourke KI; Besser TE; Miller MW; Cline TF; Spraker TR; Jenny AL; Wild MA; Zebarth GL, Williams ES. PrP genotypes of captive and free-ranging Rocky Mountain elk (*Cervus elaphus nelsoni*) with chronic wasting disease. *J Gen Virol*, 1999, 80, 2765-2769.

[116] Hamir AN; Kunkle RA; Cutlip RC; Miller JM; Williams ES, Richt JA. Transmission of chronic wasting disease of mule deer to Suffolk sheep following intracerebral inoculation. *J Vet Diagn Invest*, 2006, 18, 558-565.

[117] Miller M, Williams E. Prion disease: horizontal prion transmission in mule deer. *Nature* 2003, 425, 35-36.

[118] Miller MW; Williams ES; Hobbs NT, Wolfe LL. Environmental sources of prion transmission in mule deer. *Emerg Infect Dis* 2004, 10, 1003-1006.

[119] Marsh RF, Hadlow WJ. Transmissible mink encephalopathy. *Rev Sci Tech*, 1992, 11, 539-550.

In: Molecular Aspects of Infectious Diseases
Editor: Mazen T. Saleh
ISBN: 978-1-61728-690-2

Chapter X

Diagnostic Microbiology

***Kenneth L. Muldrew*[1] *and Yi-Wei Tang*[2]**
Molecular Genetics Laboratory, University of North Carolina at Chapel Hill Hospital, Anderson Building, Room 1031, East Wing, 101 Manning Drive, Chapel Hill, NC 27514[1]
Molecular Infectious Diseases Laboratory, Vanderbilt University Medical Center, 4605 TVC, 1161 21st Avenue South, Nashville, TN 37232-5310, Telephone: (615) 322-0126, FAX: (615) 343-8420, E-mail: yiwei.tang@vanderbilt.edu[2]

Abstract

This chapter describes methods that are used in the medical microbiology laboratory to diagnose infectious diseases. The techniques can be traditional (microscopic examination, culture on media, biochemical tests) or modern such as in vitro nucleic acid amplification tests utilizing PCR to detect/quantify organismal DNA or RNA. Further, the patient's response to an infection can be assessed through detection of the production of antibodies that are specific for a particular microbial pathogen using serology assays such as enzyme immunoassays. Physicians typically utilize all three categories of testing in order to arrive at the correct diagnosis.

Introduction

Diagnostic microbiology evaluates whether suspected pathogenic microorganisms are present in specimens isolated from a several sources including human beings, animals, and environment. Why is it important to get the correct microbial organism identification in diagnostic microbiology? Many organisms do not respond to the same antibiotics and/or treatments. Knowing the "bug" can help the Physician treat with the right "drug" that will cure the patient. For example, a skin cellulitis due to *Streptococcus pyogenes* would be successfully treated with the well known drug penicillin G. If a physician just assumed that every cellulitis was *Streptococcus pyogenes* and utilized penicillin, many patients would not get adequately treated. Cellulitis can also be caused by methicillin resistant *staphylococcus aureus* (MRSA) which is inherently resistant to penicillins. In this case, if the physician treated with penicillin, the patient would likely get a generalized infection, bacteremia, sepsis, and possible even die from the disease without the correct antibiotic (vancomycin).

In the preceding chapters various human pathogens, the human defense against these pathogens, the interaction of the host and microorganism, and new or changing infections have been described. This chapter describes how our understanding of these principles can be used effectively to diagnose microbial infections in human. We will discuss traditional methods of microscopic and culture examination along with other phenotypic (observable characteristics originating from gene expression) of these organisms. A discussion of other phenotypic characteristics that are effective tools for diagnosis will include organism growth characteristics in different media, growth after exposure to different temperatures or compounds, biochemical profiles, biochemical composition, and antigen production.

Testing for the patient's immune responses such as species-specific antibody production has become increasingly important for identification of infections in which traditional methods have been inadequate. Methods such as enzyme immunoassay (EIA), immunoblotting, and agglutination illustrate how exploitation of these immune system responses can be used to identify and treat infectious diseases.

With the advent of the polymerase chain reaction (PCR) in the 1984, the genotype of an organism has been extensively utilized to identify the species and determinants of virulence and antibiotic resistance. These modern genotypic methods have drastically reduced the test turnaround time (1-3 days down to 1-3 hours) it takes to identify organisms in the clinical laboratory which in turn, has helped physicians provide better patient care resulting in improved outcomes. This chapter will discuss various methods used in genotypic analysis including nucleic acid extraction, in vitro nucleic acid amplification techniques and post-amplification analyses for detection and identification of microorganisms. Other techniques including non-amplification probe techniques and multiple pathogen detection in one sample (multiplex assays) will also be described. Finally, large scale nucleic acid analysis using high throughput DNA sequencing and/or nucleic acid arrays hold incredible promise as well as challenges for the future of diagnostic microbiology.

Direct Examination and Antigen Testing

Diagnostic microbiology begins with sending a sample (i.e. tissue, pleural fluid, abdominal fluid, pus, sinus contents, cerebrospinal fluid, blood, sputum, etc.) to the laboratory for analysis. The sample is initially stained (usually the Gram-stain) and applied (plated) to various media, depending on the organisms suspected.

Microscopic Examination

Direct examination by microscopy of a sample for organisms is as old as microbiology itself. Aton Van Leeuwenhoek (1632-1723) - father of microbiology and inventor of the microscope- was the first to perform direct sample examination to describe microorganisms (or animalcules as he called them). This diagnostic tool is still just as important today in evaluating a specimen for the presence of pathogenic microbes.

The simplest procedure for direct examination is the wet preparation in which a portion of sample (or organism) is suspended in saline and applied to a glass slide and examined. This is useful for detection of yeast, bacterial motility, and parasites such as *Trichomonas vaginalis* which is the cause of a sexually transmitted disease. An unusual but important organism causing meningitis in children is *Listeria monocytogenes* which has a characteristic "tumbling" motility to its small rod shaped, Gram positive cell, when seen in a wet preparation.

Common Stain Methods

Several types of stains have been applied to enhance detection of particular types of organisms and allow for some level of differentiation of classes of organisms. The Gram stain stains certain types of bacteria depending on the extent of peptidoglycan in their cell wall. The Gram stain has three essential components:

(i) methylene blue stain and iodine
(ii) wash/decolorizing,
(iii) safranin counter stain.

The Gram positive rods have a thick peptidoglycan layer which stains intensely with the methylene blue and the Gram negative rods have a thin peptidoglycan layer resulting in minimal staining that can be washed out. After washing the methylene blue out, Gram negative organisms are stained red by the safranin counter stain that binds lipid cell wall. Gram positive organisms can usually be treated with antibiotics that disrupt the peptidoglycan cell wall while Gram negative organisms are treated with agents including 3rd and 4th generation cephalosporins, aminoglycosides, and pipercillin/tazobactam.

Other stains and their applications are illustrated in Table 1 and examples are shown in Figure 10.1. Figure 10.1A is a Wright's stained peripheral blood smear from a patient with a blood parasitic infection by *Babesia sp.* Ring forms are seen within the red blood cells and the black arrow points to an "X"-shaped or "Maltese cross"-shaped parasite which is diagnostic of infection by this organism. Parasitology diagnostics focuses on morphologic assessment and to a lesser extent antigen testing as opposed to biochemical and culture methods. Figure 10.1B is a Methenamine silver tissue stain of sinus contents from a patient with a filamentous fungal sinusitis due to *Aspergillus fumigatus.* Figure 10.1C shows a lactophenol cotton blue stain of an *Aspergillus fumigatus* colony isolated on Sabouraud Dextrose Agar (also see Figure 10.2A) and it produces the characteristic fruiting structure seen with this organism that is also visualized on the direct preparation silver stain (Figure 10.2B). Figure 10.1D shows a partial acid-fast stain of the branching rod known as *Nocardia asteroides.* It can also be seen as a branching Gram positive rod on the Gram stain. *Mycobacterium sp.* can be stained with Auramine-Rhodamine (Figure 10.1E) which is a fluorescent molecule that binds to the thick mycolic acid cell wall of Mycobacteria and can be visualized with a fluorescent microscope. It increases the sensitivity of detecting low numbers of organisms in a direct sputum smear and allows the technologist to screen a large area of the microscopic slide in a short period of time. A fluorescent antibody stain for *Chlamydia sp.* is depicted in Figure 10.1F. The antibody is applied to a human cell monolayer that has been previously inoculated with patient specimen. If organism is present, the intense staining can be visualized in the fluorescent microscope. This same technique is also applied to detecting viruses in patient's samples.

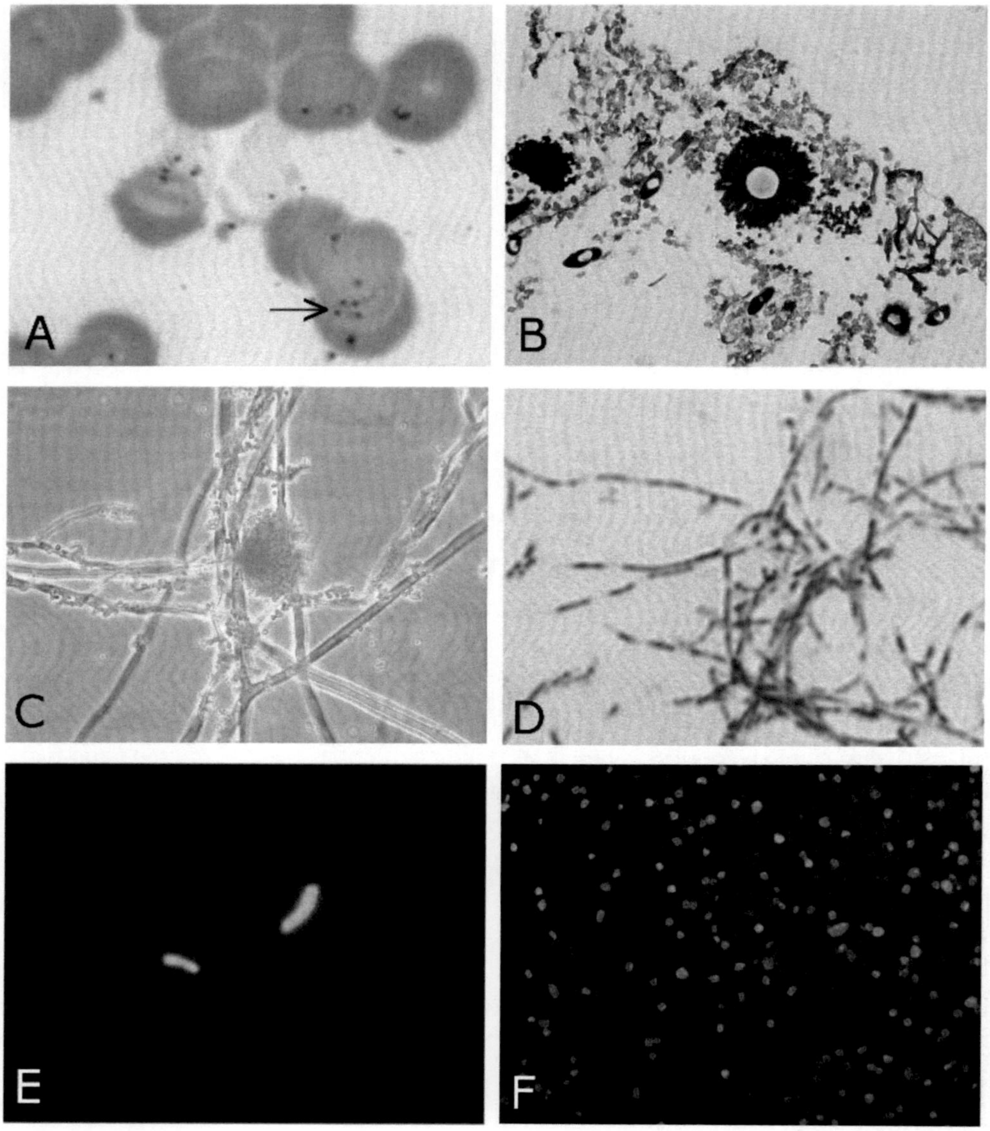

Figure 10.1. Stain techniques used in diagnostic microbiology. Figure 1A. Wright's stained peripheral blood smear from a patient with Babesia sp. infection. Ring forms and an "X"-shaped or "Maltese cross"-shaped parasite are present which is diagnostic of the infection. Figure 1B is a Methenamine silver tissue stain of sinus contents from a patient with Aspergillus fumigatus sinusitis. Figure 1C Lactophenol cotton blue stain of an Aspergillus fumigatus colony present on Sabouraud Dextrose Agar (also see Figure 3A). Figure 1D Partial acid-fast stain of the branching rod Nocardia asteroides. Figure 1E shows a positive Mycobacterium sp. Auramine-Rhodamine stain which is a fluorescent molecule that binds to the thick mycolic acid cell wall of Mycobacteria and can be visualized with a fluorescent microscope. Figure 1F shows a fluorescent antibody stain for Chlamydia sp. that has been applied to a human cell monolayer that has been previously infected. Viral fluorescent antibody stains would have a similar appearance.

Table 10.1. Stains and their uses.

Stain Type	Organisms Detected	Sample Type	Utility/drawbacks
Gram Stain	Most Bacteria Yeast Other fungi	tissue body fluids blood cultures agar cultures broth cultures	Routine Screen for bacteria and fungi on a variety of specimen types and primary culture material; inexpensive, fast; cells can be affected by antibiotic treatment; Over and understaining can occur secondary to variations in technique
Acridine Orange	Bacteria Mycoplasma sp.	tissue body fluids blood cultures agar cultures broth cultures	Differential staining of bacterial & Mycoplasma; does not need a bacterial peptidoglycan cell wall; must have a fluorescent microscope
Wright's	Bacteria Yeast Other Fungi Blood parasites	peripheral blood bone marrow	Fast and accurate; only shows morphology; does not provide bacterial cell differentiation (i.e., Gram positive vs. negative)
Acid-Fast	Mycobacteria sp. (Acid-Fast Bacilli-AFB)	tissue body fluids blood cultures agar cultures broth cultures	Direct examination of sputum specimens; slow scanning of slide; does not differentiate between Mycobacterium tuberculosis and other less pathogenic Mycobacterial sp.
Auramine-Rhodomine	Mycobacteria sp. (Acid-Fast Bacilli-AFB)	tissue body fluids blood cultures agar cultures broth cultures	Direct specimen analysis; increases sensitivity for detecting AFB over Acid-Fast Stain; rapid scanning of slide; need fluorescent microscope
Modified Acid-Fast	Nocardia sp.	tissue body fluids blood cultures agar cultures broth cultures	Direct specimen analysis; increases sensitivity for detecting; relatively specific for Nocardia sp.
	Cryptosporidia Isospora Cyclospora	Stool	Fairly rapid; multiple steps
Periodic Acid Schiff (PAS)	H. pylori	formalin fixed gastrointestina l biopsy tissue	Fair sensitivity; slow-(1-2 days tissue processing time)
Calcofluor White	Fungi Pneumocystis jirovici	tissue body fluids blood cultures agar cultures broth cultures	Direct specimen analysis; increases sensitivity/specificity for detecting fungi over Gram-stain; genus can be determined by experienced users rapid scanning of slide; need fluorescent microscope
Lactopheonol Cotton Blue	Fungi	culture colony	Rapid and good for morphological examination of primary cultures
Methenamine Silver	Histoplasma sp. Other fungi	formalin fixed tissue sections body fluids	Good for morphology; slow-(1-2 days tissue processing time)

Biochemical and Antigen Testing

Like all other organisms, microorganisms undergo metabolism and produce byproducts that can be detected using color indicators and substrates. For Enterobacteriaceae, sugar metabolism, indole reaction, and urease are important reactions in differentiating between species. Anaerobes produce many organic acids because of anaerobic metabolism and these can be used to establish patterns for specific species.

Figure 10.2A demonstrates the Triple Sugar Iron (TSA) slant that tests organisms for their ability to ferment glucose and lactose as well as to produce gas and hydrogen sulfide (H_2S). *Pseudomonas sp.* are non-fermenters (2nd tube from right). Tubes 3 through 6 show 4 fermenting Gram negative rods that produce acid from sugar metabolism and thus produce a yellow color in the media. The black color is an iron precipitate that forms from H_2S production. If *Proteus* did not produce H_2S, the tube would look yellow (tube 6). *Salmonella typhi* shows fermentation but has a limited amount of H_2S production at the top of the tube which is characteristic of this organism.

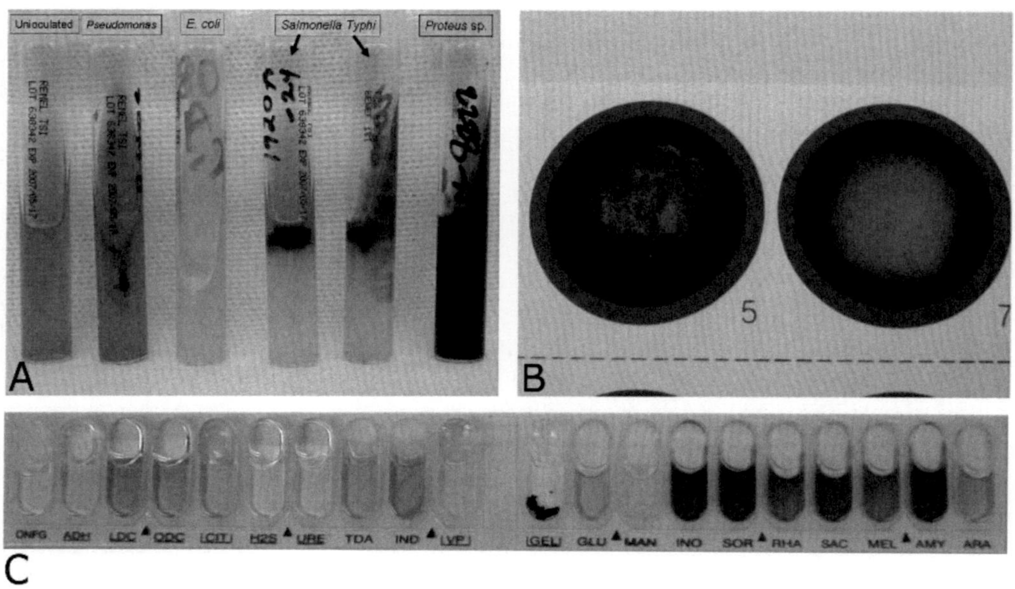

Figure 10.2. Biochemical and antigen testing methods. Figure 2A demonstrates the Triple Sugar Iron (TSA) slant. Pseudomonas sp. are non-fermenters (2nd tube from right). Tubes 3 through 6 show 4 fermenters producing a yellow color. The black color is an iron precipitate that forms from H2S production. Proteus (tube 6) is H2S positive and a fermenter. Salmonella typhi shows fermentation but has a limited amount of H2S production (tubes 4 &5) the characteristic "typhi smile". Figure 2C API 20 E biochemical test strip showing the reactions that occur with E. coli. This same E. coli organism was tested for the presence of the O1H757 antigen (Figure 2B) using a latex agglutination antibody test. When the antigen is present, the beads clump together (left circle); right circle is the negative control without organism.

Different biochemical activities can produce obvious phenotypes which allow the medical microbiologist to differentiate one species from another. Figure 10.2C is a biochemical test panel of dry reagents that are incubated with a solution of pure organism which produces characteristic color changes. The pattern (or biotype) is specific for a certain species, in this case- *E. coli.* This same *E. coli* organism was tested for the presence of the

O1H757 antigen (Figure 10.2B) which is a cell surface antigen that is associated with the pathogenic *E. coli* that causes hemolytic-uremic syndrome. A small amount of the organism is incubated with a reagent containing anti-O1H757 antibodies bound to tiny plastic beads. When the antigen is present, the beads clump together (left circle) which is easily differentiated from a negative control (right circle) in which the beads stay in solution. *Staphylococcus aureus* is differentiated from other *Staphylococcal sp.* in the same way by the use of the coagulase test. Matrix-assisted laser desorption ionization-time-of-flight mass spectrometry (MALDI-TOF MS) has recently been reported for rapid and accurate microbial species identification (Figure 10.3). A system called BioTyper based on the MALDI-TOF MS is commercially available from Bruker Daltonics, (Leipzig, Germany).

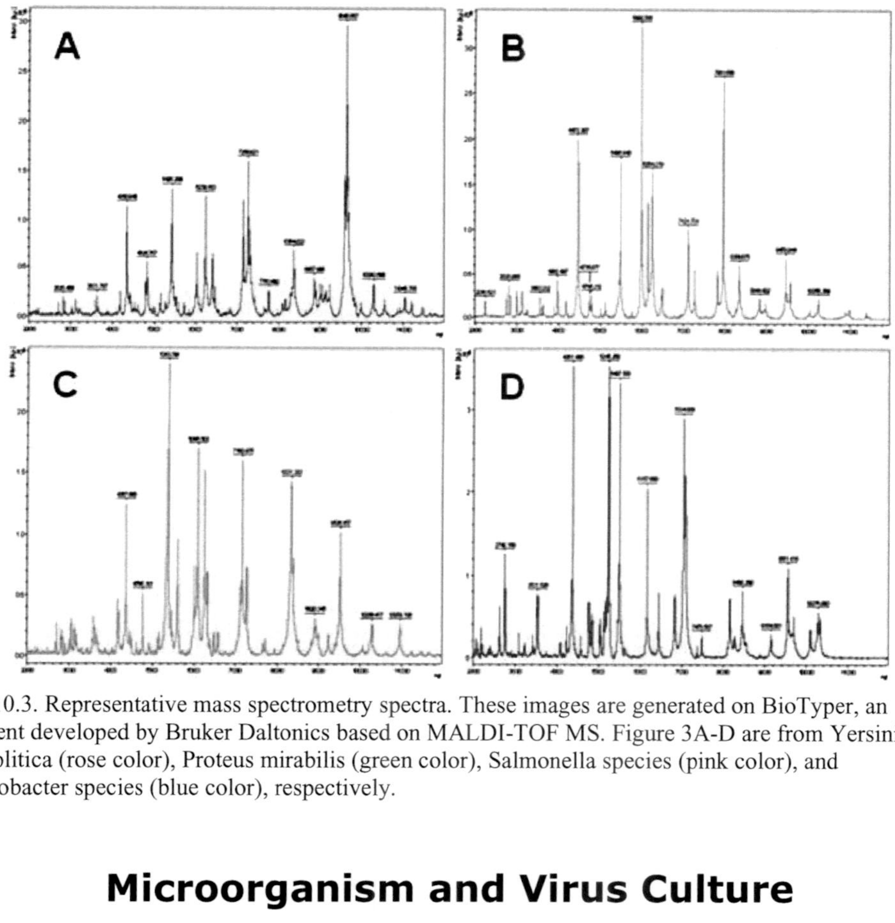

Figure 10.3. Representative mass spectrometry spectra. These images are generated on BioTyper, an instrument developed by Bruker Daltonics based on MALDI-TOF MS. Figure 3A-D are from Yersinia enterocolitica (rose color), Proteus mirabilis (green color), Salmonella species (pink color), and Campylobacter species (blue color), respectively.

Microorganism and Virus Culture

Primary Alive Culture

Once a direct Gram stain is performed, the sample will be applied to various media for culture. Table 2 describes various media and their uses. Multiple media types are utilized on each sample in order to provide differential growth information on different substrates. For example, if the organism grows on MacConkey's agar and not on the Colistin Naldixic Acid agar (CNA), then the organism is a Gram negative. The reverse situation would be a Gram positive organism. When the Gram stain of the direct specimen is negative, this can provide more information. A subsequent Gram stain of the colony and further biochemical/antigen testing can be utilized to arrive at the final identification.

Table 10.2. Examples of Media and their Uses.

Media Type	Organisms Detected	Description
5% Sheep Blood Agar (SBA)	Most Bacteria Yeast Other fungi Rapid growing Mycobacteria	This agar is a general all purpose non-selective agar that grows most aerobic organisms but not viruses, anaerobes or slow growing Mycobacteria
Chocolate Agar	Most Bacteria Yeast Other fungi Rapid growing Mycobacteria.	Similar to SBA this is also a general all purpose non-selective agar that grows most aerobic organisms but not viruses, anaerobes or slow growing Mycobacteria. It will also grow some of the more fastidious organisms like Haemophilus that needs extra nutrients that are provided in the formulation
CDC blood agar	Anaerobic bacteria Streptococci	General purpose anaerobic media that must be incubated in an anaerobic chamber to grow anaerobes
MacConkey Agar	Gram negatives	Gram negative selective agar that is differential for lactose fermentation by turning the media pink due to metabolism of lactose
Sabouraud-Dextrose Agar	Yeast Other fungi	Fungal- selective agar that is optimized for yeast and filamentous fungi and has antibiotics that inhibit the growth of bacteria
Lowenstein-Jensen Agar	Mycobacteria	Selective for the growth of Mycobacteria. Contains long chain lipids necessary for the formation of the large Mycobacterial cell wall
Thioglycolate Broth	General Purpose non-selective broth (except for viruses)	Non-selective agar that can grow both anaerobes and aerobes. Very sensitive but can grow lots of anything
Kiryby-Bayeur Agar	Aerobic bacteria Facultatively anaerobic bacteria	Utilized for antibiotic susceptibility testing of aerobes and facultative anaerobes
RhMK cell monolayer	Influenza Virus Parainfluenza Virus Enteroviruses	Primary Rhesus Monkey Kidney cells in a monolayer in a glass test tube
Hep2 cell monolayer	Adenovirus Respiratory Syncytial Virus Cocksackie B virus	Cells from a Human squamous cell carcinoma of the larynx in a monolayer in a glass test tube
MRC-5 cell monolayer	Cytomegalovirus Varicella Zoster Virus Rhinovirus Herpes Simplex Virus	Human Embryonic Lung Fibroblasts in a monolayer in a glass test tube

Figure 10.4A demonstrates a fungal media (Sabouraud Dextrose Agar fungal culture of *Aspergillus fumigatus*. This media contains antibiotics (to inhibit bacteria) and nutrients that enhance the growth of fungi. Figure 10.4B and 4C show cultures of a Gram negative rod (*E. coli*) on a non-selective sheep's blood agar plate and a selective and differential Gram negative plate (MacConkey's Agar). The blood plate will grow most aerobic and facultative

anaerobes and is thus good for isolation. MacConkey's agar is "selective" in that it will only grow Gram negative organisms (has antibiotics against Gram positives).

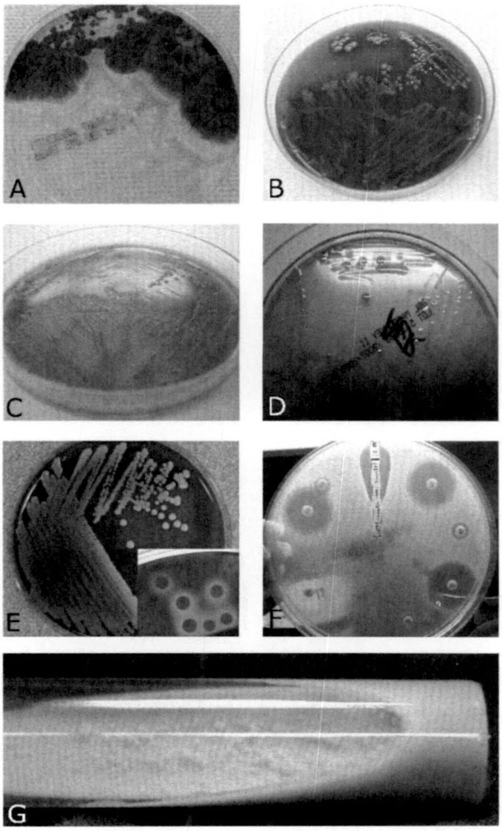

Figure 10.4. Agar culture methods. Figure 3A. Fungal media (Sabouraud Dextrose Agar) fungal culture of Aspergillus fumigatus. Figure 3B and 3C Gram negative rod (E. coli) on a non-selective sheep's blood agar plate and a selective and differential Gram negative plate (MacConkey's Agar). Figure 3D shows a non-fermenter. (Figure 3E) Colistin Naldixic Acid Agar (CNA) growing Staphylococcus aureus (SA). Inset shows beta hemolysis (zone of clearing) of SA. Figure 3F. Kirby-Bauer agar plate for antibiotic resistance testing of organisms such as MRSA. Mycobacterium tuberculosis (Figure 3G) specialized media (Lowenstein-Jensen Agar).

It is also a "differential" media in that it will turn pink when organisms ferment lactose (pH color indicator) and will remain colorless (Figure 10.4D) when it is a non-fermenter. Colistin Naldixic Acid Agar (CNA) is another "selective" agar because it will only grow Gram positive organisms (Figure 10.4E) such as *Staphylococcus aureus* (SA). Because it is a "Blood-Based" agar, it can reveal the characteristic beta hemolysis (zone of clearing) of SA (Figure 10.4E, inset). Agar based methods are also utilized to detect antibiotic resistance (Figure 10.4F) of organisms like Methicillin-Resistant Staphylococcus Aureus (MRSA). A zone of clearing demonstrates that the organism is susceptible to the antibiotic and no (or, a small zone) zone, indicates the organism is resistant. Antibiotic susceptibility patterns can also help in differentiating the species of an organism. *Mycobacterium tuberculosis* (Figure 10.4G) has unusual nutritional requirements (long chain lipids) and therefore has its own specialized media (Lowenstein-Jensen Agar).

Culture using Living Cells

Many viruses and *Chlamydiae* can be isolated and identified using eukaryotic cell cultures. The procedure begins by inoculation and incubation of the sample (frequently respiratory) on several types of cell monolayer cultures (Table 2). When organism is present in the sample, the cell monolayer becomes infected and the cells change their shape, become distorted, and eventually the cells will lyse. The cell shapes and sizes and patterns of cell destruction can be characteristic for certain viruses. To confirm the presence of a specific virus, fluorescent antibodies are used that will detect only that virus (e.g. RSV antibody). Not all viruses can be isolated by these methods, but it is very sensitive in that it takes just one viral particle to infect the cell monolayer and propagate the infection throughout.

Cell culture methods systems used for virus isolation are listed in Table 2. After inoculation, a virus can initially be identified and differentiated by the type of viral cytopathic effect (CPE) seen, the speed with which the CPE begins after inoculation, and the type of cell culture within which the virus replicates (Figure 10.5). Enteroviruses for example, will produce a characteristic CPE in rhesus monkey kidney cells. Of the enteroviruses, echovirus will show CPE in 4 days while poliovirus CPE is present within 1 day after inoculation. Multiple different cell cultures will be inoculated for each sample to increase the yield of viral culture as well as help differentiate the type of virus isolated.

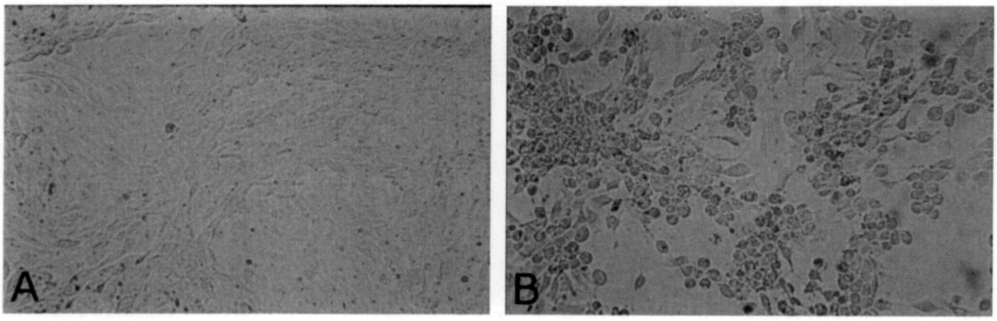

Figure 10.5. Viral cytopathic effect (CPE). Figure 4A shows a normal, uninfected rhesus monkey kidney cell line. After inoculation and inoculation of the cell line with polio virus, the characteristic CPE is demonstrated in Figure 4B.

A variant of the cell culture technique which detects antigens is Shell vial isolation. This can be used to detect viruses and *Chlamydiae*. This method is rapid compared to traditional culture which can take up to three weeks for isolation. Glass vials contain coverslips that have a cell monolayer that is bathed in media. When the monolayer is confluent over the coverslip, then the sample is added, incubated centrifuged and incubated. The coverslip in the vial is then incubated with fluorescein-conjugated virus-specific antibody. The stained coverslip is then removed and examined under a fluorescence microscope. Detection of cytomegalovirus (CMV) infections is greatly enhanced by the 1-2 day shell vial technique, since CMV culture can take up to three weeks to reveal the characteristic CPE. A genetically altered cell line, the enzyme-linked virus inducible system (ELVIS, Diagnostic Hybrids, Inc., Athens, OH), is a genetically manipulated cell line that has been used to rapidly detect herpes simplex viruses by cell culture. These types of Transgenic and co-cultured cell lines remove the necessity for multiple vials and also increase the sensitivity for detecting multiple Enteroviruses. Mixed cell lines are available that contain host factors that enhance binding and entry of some enteroviruses (Super E-Mix) and these are commercially available. (Diagnostic Hybrids, Inc.).

Immunological Diagnosis

To examine patient's response to organisms, several methods are utilized. Serology assays evaluate the presence of IgM and IgG antibodies that are made in response to microorganisms. IgM is an indicator of a new and ongoing infection. When acute and convalescent titers are obtained, A four fold increase indicates recent, active infection. Detection of specific anti-organism antibodies in the patient is useful for organisms that are not easily cultivated such as HIV, Hepatitis viruses (A, B, C), Mycoplasma sp., Chlamydia sp. and Syphilis. Cultivatable viruses are also tested by this method (HSV, VZV, CMV, and EBV). This method can also be used to detect organism antigens and is similar in nature to the E. coli O157H7 test depicted in Figure 5B. Detuned antibody assays evaluate the avidity of an antibody by dilution or treating tested serum specimens; this has been helpful in making the distinction between recent and remote infections. This technique has been used primarily in HIV clinical research for investigations and epidemiology surveillance. In immunocompromised hosts, serology may be limited due to the likely hood of an inability to mount an effective immune response.

Immunofluorescence Methods

Immunofluorescence methods have been extensively used in the past for detection of specific antibodies to microorganisms and continue to have several applications today. The common procedure used is indirect immunofluorescence. When the particular antibody is in the serum specimen it binds with a microorganism-specific antigen bound to a pre-fixed glass slide. The slide is washed and then a fluorescein-conjugated antibody directed against the test serum species is applied. After subsequent wash steps, fluorescence is detected by a fluorescence microscope. Several viral infections are detected by primarily by this method such as measles virus and Epstein-Barr virus by many clinical laboratories. There are limitations to this technique in that it is labor intensive and the reading of the fluorescence is subjective. Other more objective techniques such as enzyme immunoassays (EIA) are gradually replacing this method.

Enzyme Immunoassays

The most widely used diagnostic method in the clinical serology laboratory is the enzyme immunoassay. The most frequently used method is the solid-phase EIA technique. This uses plastic microtiter plates or beads to which antigens passively bound. A patient's serum specimen is applied and if present, the microorganism-specific antibody binds to the antigen. An enzyme-labeled immunoglobulin specific for the test serum antibody is added after a washing step. A second was step is performed and then a chromogenic enzyme substrate is applied, producing a colored product that is directly proportional to the amount of specific antibody in the patient's sample. Utilization of the EIA technique in the diagnosis of viral and ricketsial infections has greatly increased in recent years. It is frequently used to screen a patient's immune status, for viruses such as rubella. This assay is also used to screen for various other viral infections such as human T-cell lymphotrophic virus, hepatitis B virus (HBV), HIV, and hepatitis C virus (HCV).

Immunoblotting

In direct test formats, serologic test specificity is largely defined by the antigen used to capture the antibody of interest. One of the most specific serologic methods is the immunoblot (or Western blot). First, protein antigen is separated by sodium dodecyl sulfate polyacrylamide gels which is then transferred by electrophoresis to a nitrocellulose membrane. Free-protein-binding sites are blocked on the membrane and test serum is added and specific antibodies are bound to the antigen and detected. Multiple organism specific antigens versus cross-reactive proteins can be differentiated resulting in a very specific assay for a particular microorganism. Immunoblotting is widely utilized for confirming infection with *B. burgdorferi*, HCV, and HIV after the patient initially tests positive for the particular screening test. An example and description of HIV immunoblotting is presented in Figure 10.6.

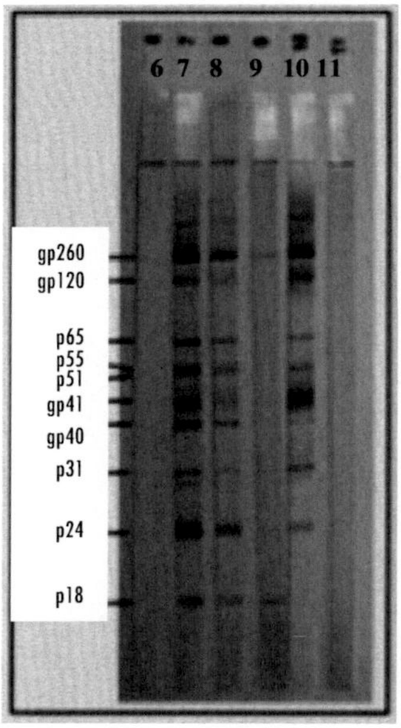

Figure 10.6. HIV immunoblot. This image shows an immunoblotting test for HIV. Strip 6 is a negative control. Strips 7-8 are high and low positive controls respectively. Strips 9-11 are indeterminate, positive and negative patient results respectively.

Other Serologic Techniques

Complement fixation and agglutination assays are other serologic techniques to assess the humoral response. Complement fixation is still used in the serodiagnosis of several fungal infections. It is a very labor intensive assay and does not provide optimal sensitivity. It will be gradually replaced over time. Agglutination testing uses polystyrene (latex) particles coated with organism antigen. When exposed to a test sample, specific microorganism immunoglobulin will bind to the beads and cause them to aggregate which can be seen on the test card. These tests are easy to perform and take little time (15-30 minutes). This test has been used in the clinical laboratory for rapid plasma regain quantitation that is used to monitor syphilis treatment and identifying rubella virus antibodies.

Newer serologic techniques have been recently developed. A single reaction, multianalyte antibody panel profile (xMAP) allows differentiation and detection of multiple antibodies within a patient's specimen using xMAP multiplexed technology (Luminex, Austin, TX). This expanded information can provide a better differential diagnosis to assess for infectious diseases.

Molecular Diagnosis

This section describes the utilization of nucleic acid analysis to identify, quantify, and type microorganisms. We will first mention nucleic acid extraction which is important to many of these assays. Non-amplified probe methods are used as direct ways to assess organism in a sample or culture without nucleic acid amplification. Amplification methods will be described including signal amplification and target amplification techniques. Alternative analyses for detection, identification, and typing of microorganisms include sequencing, reverse hybridization, and liquid suspension array analysis will also be discussed. Finally, large scale nucleic acid analysis using high throughput DNA sequencing and/or nucleic acid arrays are important because these techniques are at the cutting edge of a developing science of whole genome analysis for the use in the clinical laboratory.

Nucleic Acid Extraction

One of the more important, but often overlooked steps in molecular diagnostics is the nucleic acid extraction. This part of an assay gives us the target to employ in further downstream analyses such as PCR. A good nucleic acid extraction method is comprised of efficient target recovery, establishment of the integrity of nucleic acid targets, optimal removal of amplification inhibitors, elimination of components which affect other enzymatic substrates, and sterilization of potentially hazardous organisms. It begins with organism/cell lysis using reagents such as detergents and/or heat, and then protein digestion using a protease. Once liberated, the nucleic acid is differentially isolated from the other macromolecules (lipids and proteins) by adding an ethanol/salt solution which precipitates the nucleic acid. The precipitate can be resuspended in buffer and utilized for other downstream applications.

Two other variations of this procedure are phenol-chloroform and silica column extraction. In the phenol-chloroform extraction, proteins and lipids are removed into the organic layer and nucleic acids are retained in the aqueous layer and can be isolated and used for further downstream applications. Finally, Silica column extraction employs the use of silica or glass membranes in the bottom of a filter cup. The sample is exposed to chaotropic salts (e.g. NaI) which have two affects. They destabilize proteins and other macromolecules by affecting hydrogen bonding and increase the affinity of Nucleic acids for the silica membrane promoting binding. Upon washing with buffers to remove unwanted residual macromolecules, the NA on the column can be eluted in water or TE buffer and then used for downstream analyses such as PCR.

A variety of manual and automated kits and methods have been developed for nucleic acid extraction from a variety of patient specimen types. QIAamp MinElute virus kits (Qiagen Inc., Valencia, CA) can efficiently extract total viral nucleic acids from respiratory specimens and cerebrospinal fluid. The easyMAG platform (bioMérieux, Durham, NC) has become more widely used, because it provides good quality nucleic acids with a high yield, is automated, and flexible.

Non-Amplified Probe Methods

Non-amplified probe methods do not employ PCR and thus have no risk of contamination. An example is the peptide nucleic acid (PNA) probe which can more easily enter cells because of its increased hydrophobicity. It also binds more tightly to the target because of the uncharged peptide backbone (compared to regular DNA probes). Bases are attached to the peptide backbone to provide specificity for the organism tested. These probes have a fluorescent label that can be detected by a fluorescent microscope. Probes have been designed to detect *Staphylococcus aureus* and *Candida sp.* from blood culture bottles and *Mycobacterium tuberculosis* from direct sputum smears.

Amplification Methods

Signal Amplification

Signal amplification methods, like the Digene Human Papilloma Virus test, incorporate the generation of multiple signaling molecules, increasing the sensitivity of detection over direct probe methods. Since these methods do not amplify DNA by PCR, there is no risk of amplicon contamination. The HPV assay is called the hybrid capture method in which an RNA probe binds to the viral DNA that is in turn, bound by a "capture antibody" that is immobilized to the solid surface of the reaction tube. An enzyme-labeled antibody is added that binds the hybrid and a substrate is added which then is turned into a chemiluminescent signal. This assay is very sensitive and provides multi-genotype information. This format has also been used to detect other organisms such as *Chlamydia trachomatis.*

PCR-Target Amplification

PCR has revolutionized laboratory detection, typing, and identification of microorganisms (Figure 10.7). The method hinges on a very important enzyme *Taq* DNA polymerase that was discovered in the bacteria *Thermophilus aquaticus* (thus the name Taq) which grows well in very hot environments (100+ °C). This enzyme is utilized in a temperature cycling set of steps along with primers (small complementary pieces of DNA that bind the target of interest) to amplify a small DNA template that is specific for the organism (Figure 10.7). Initially the reaction vessel is heated and the *recA* bacterial DNA strands separate. The temperature is decreased to 60 degrees Celsius allowing the *recA*-specific primers to bind on either single stranded portion of the target. These primers bind in a 5' to 3' configuration that is complementary to the *recA* bacterial sequence. The reaction is heated to 72 degrees Celsius and the *Taq* polymerase comes along and binds to the primer region and extends the sequences (on both strands) to produce two new DNA strands. This is done 30 times to create 2^n copies of template (n=number of cycles) $=2^{30}=1.1 \times 10^9$ (or 1.1 billion) copies of *recA* template in the reaction.

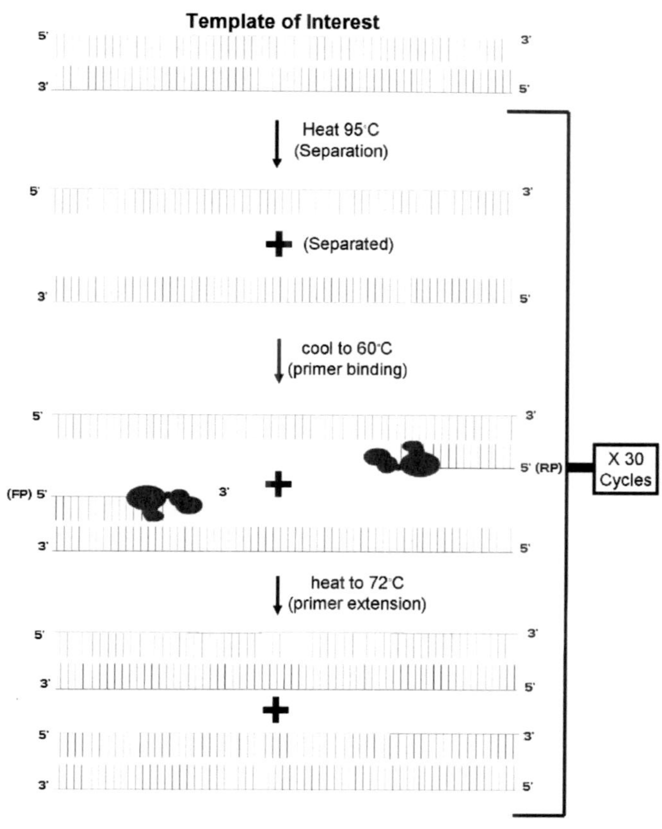

Figure 10.7. Polymerase chain reaction. This figure demonstrates the steps of the PCR in which microbial DNA strands are separated by heating to 95 degrees Celsius and then colling. As the reaction cools, the sequence-specific primers will bind and then Taq polymerase binds to this region. The temperature is increased to 72 degrees Celsius and the primer is extended by the Taq polymerase to create two new daughter strands. These same three steps are repeated 30 times, generating millions of template DNA copies.

Other methods that utilize target amplification include nucleic acid sequence-based amplification, strand displacement amplification, transcription-mediated amplification, and helicase-dependent amplification. Products using these techniques are available from Becton, Dickson and Company (Sparks, MD), GenProbe (San Diego, CA), BioHelix (Beverly, MA) and bioMerieux (Durham, NC). These products can provide quantitative data for *Clostridium difficile*, HIV, *M. tuberculosis*, *Neisseria gonorrhoeae*, MRSA, *Chlamydia trachomatis*, enteroviruses, and HCV. Target amplification methods are very sensitive and can detect very small amounts of microorganism nucleic acid within a patient specimen. As a result of this sensitivity, there is a potential for contamination of a subsequently analyzed sample with the patient's amplified material. If proper procedures are followed with separation of pre and post amplified laboratory sections, this potential problem can be virtually eliminated.

Real Time PCR

Real-time PCR is used in the detection and quantification of a variety of viruses and difficult to cultivate organisms such as HIV, CMV, VZV, EBV, BKV, Influenza Virus, *Chlamydia sp.,* and *Mycoplasma sp.* A variation of the basic PCR method, rt-PCR incorporates an additional short DNA sequence that also contains a fluorescently labeled molecule on the 5' end and a quencher molecule on the 3' end. This sequence is specific for an area in between the primers. When exposed to a laser or diode light source in the instrument, the fluorescent molecule is excited. Because there is a quencher molecule on the probe, this energy is dissipated by transference to the quencher molecule in a process termed fluorescence resonance energy transfer (FRET). As annealing and extension takes place, the probe binds and is "chewed up" by the 5' to 3' exonuclease of Taq polymerase. This allows the 5' fluorescent molecule to be completely disassociated from the quencher molecule and thus no FRET occurs. The energy is then released in the form of a photon which is then captured by the instrument's detector and recorded by the computer software as signal.

Another variation of this technique is to use a fluorescent probe and a quencher probe that bind to the region of interest if it is present in the sample. This binding causes a temporary elimination of a portion of the fluorescent signal and over 45 cycles this will increase in proportion to the amount of template present in the original sample. Since no probe is degraded, melt curves can be performed which provide more specific information about the organism while still retaining the ability to quantify the amount present by comparison with a standard curve of known template (see below).

Figure 10.8A shows an AB 7500 real-time TaqMan assay standard curve of log base 10 dilutions of a known template and the increasing fluorescence that is seen as the number of PCR cycles increases (bottom graph). The earlier the signal crosses the threshold line (horizontal line), the larger the amount of template that was in the original sample in the beginning of the PCR reaction. Because this is a standard curve, the exact concentrations of template are known (upper graph) and this can be plotted against the Crossing Threshold (Ct) value to get a linear plot from which unknown concentrations of template can be derived in clinical samples. Figure 10.8B shows a Light cycler melt curve analysis with two different melting temperatures. This result could be a genotyping assay (for example HSV1 vs. HSV2) in which each of the two genotypes has a slightly different sequence, which results in different melt curves and thus different melting temperatures (Tm). Not only does it show the presence of the HSV in the sample (the HSV specific sequence was amplified), but based on the Tm, you can confirm the genotype (HSV1 or HSV2).

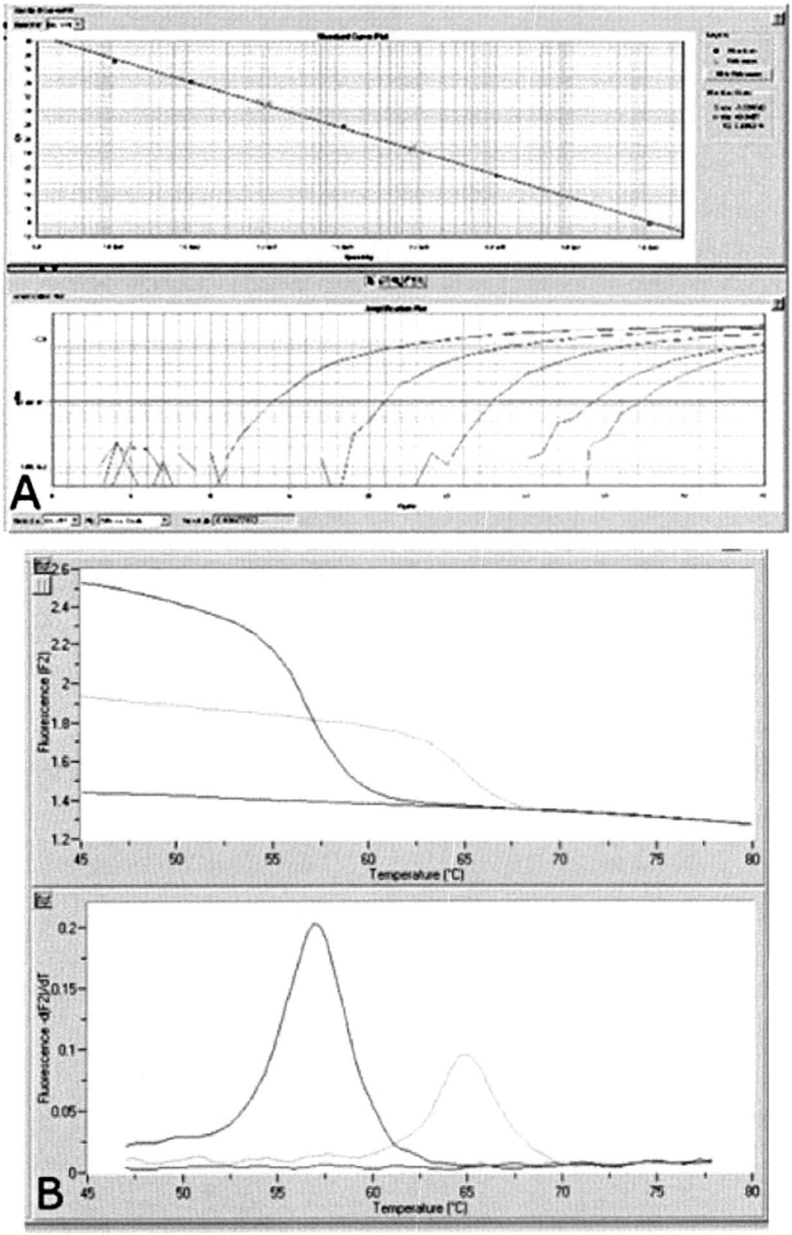

Figure 10.8. Real-time PCR. Figure 8A shows an AB 7500 real-time TaqMan assay standard curve of log base 10 dilutions of a known template and the increasing fluorescence that is seen as the number of PCR cycles increases (bottom graph). Figure 8B shows a LightCycler melt curve analysis with two different melting temperatures. This result could be a genotyping assay (for example HSV1 vs. HSV2) in which each of the two genotypes has a slightly different sequence, which results in different melt curves and thus different melting temperatures (Tm).

Real-time PCR has been expanded to have the ability to detect multiple (3-5) targets in one reaction vessel (multiplex). Multiplex assays are difficult to design because multiple primers and probes are needed for amplification and detection of different targets and these all compete and to some extent will bind with each other and can decrease the efficiency of the individual reactions. Despite design issues, numerous good assays have been developed for use in the clinical laboratory such as the IDI-MRSA and Chlamydia/Mycoplasma/ Legionella sp. assays. Several novel multiplexing procedures including target-enriched multiplexing PCR owned by Qiagen (Valencia, CA) have been invented to enhance efficacy of multiplex amplifications.

Post Amplification Analyses

After amplification of the initial template in PCR other amplification methods, effective detection methods need to be performed to evaluate for a specific microorganism. Biological amplification (culture growth) can be easily visualized on a plate or in a test tube, but after amplification the material is not visible to the naked eye. It is important that these methods be relatively simple and easy to perform, and provide the sensitivity and specificity that is important in the clinical laboratory. Thus, we need a way of visualizing the results of amplification assays, one that enhances rather than hinders the sensitivity and specificity of the assay. Amplification product identification and Detection has only recently become an established procedure in the clinical laboratory.

Colorimetric ELISA

Nucleic acid amplification assays that use microwell plate detection methods provide simplicity, high throughput and convenience. It can be automated and has excellent sensitivity that is comparable to other methods. The method has similarities with the enzyme-linked immunosorbent assays (ELISA); and this product detection method has been called the EHA (enzyme hybridization assay) or ELOSA (enzyme-linked oligosorbent assay). These methods have been developed in individual laboratories and are also commercially available (Roche Diagnostics, Indianapolis, Ind.; Millipore, Billerica, Mass.; Argene, North Massapequa, NY). Only complete kits are available for the Roche format (Amplicor) for detection of microorganisms such as HCV, HIV-1, and *C. trachomatis*. Multiplex microwell-based PCR assays are available from Prodesse, Waukesha, Wisconsin and among others include the Adenoplex (adenoviruses), Hexaplex (respiratory viruses), and Pneumoplex (bacterial causes of pneumonia). Other microwell plate methods are available from Becton Dickinson, Sparks, MD, including the ProbeTec *Neisseria gonorrhoeae* and *Chlamydia trachomatis* tests.

Reverse Hybridization

Reverse hybridization is a simple process whereby specific organism specific colorimetric probes are attached to a solid membrane and then PCR products are applied. If a certain PCR product binds to a probe, then it will generate a color change in response to the addition of a substrate. Multiple applications for this technique have been employed. It has been utilized to detect *Mycobacteria sp.* and its resistance genes, *H. pylori* resistance genes, and genotyping of HCV, HBV, and *Streptococcus pneumoniae.*

Solid Nucleic Acid Arrays

Nucleic acid arrays provide another technique to rapidly analyze a very large amount of nucleic acid information to be used in the diagnosis of infectious diseases. In DNA arrays, thousands of labeled probes can be attached to tiny individual wells or spots on a chip. Applying the processed sample, detection will occur if any of the targets are present and they will be detected in a spot that is specific for the NA probe. Multiple respiratory viruses have been detected on a single chip. There is one chip that originally detected 22,000 viral targets and the later version now detects 144,000 targets. This assay can also detect virulence factors and mechanisms of antibiotic resistance. It has already proven useful in the direct detection of antibiotic resistance in *Mycobacterium tuberculosis*. Because it takes 3-6 weeks to culture and determine resistance patterns of *Mycobacterium tuberculosis*, this 2 day test is very attractive to the clinical laboratory. For this type of testing, there is a drawback in that it does not detect in vivo expression of the virulence factors and/or resistance determinates and is one step removed from detecting the true resistance. Expression patterns may be altered such that even though the DNA sequence is present, the antibiotic resistance mechanism may not be present.

Suspension Array Analysis

The Luminex technology provides a mechanism for highly multiplex assays to be brought into the molecular pathology clinical laboratory. It uses flow cytometric principles to achieve differentiation between 100 different analytes. It works by using 100 differently shaded red beads that have short DNA segments attached- a different sequence for each bead. Each one of these sequences is labeled with a probe (Figure 10.9). Multiplex PCR is first performed to amplify targets (if present in the sample) then this is applied to a tube with the beads. Only those amplified targets will bind to the bead and be detected by the probe. Since each different bead has a different NA target and bead color, the instrument can detect which NA target is present in the sample. The most notable use of this technique in molecular pathology is a 20 virus respiratory panel that has recently been developed to analyze these viruses in one multiplex reaction.

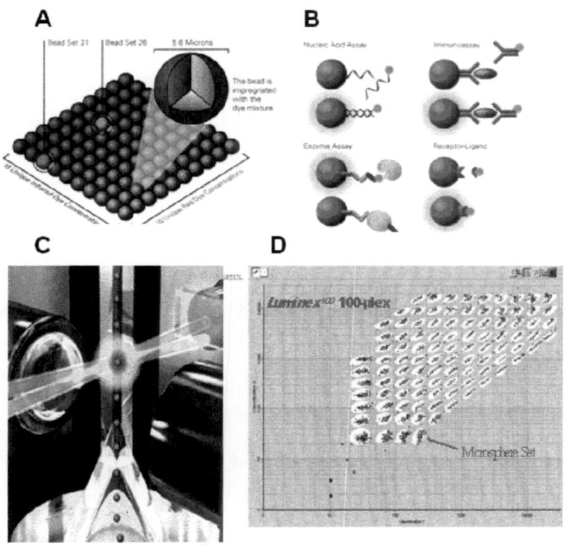

Figure 10.9. Luminex suspension array. Figure 9A shows that Liminex internally color-codes microspheres with precise concentrations of two dye yielding up to 100 distinctly colored bead sets. In Figure 9B, the microspheres can be coated with reagents specific to a particular bioassay such as antigens, antibodies, oligonucleotides, enzyme substrates, or receptors. Figure 9C shows that microspheres are interrogated as they pass single file by a red and green laser and Figure 9D indicates log plot of red fluorescence intensity versus infrared fluorescence intensity showing 100 microsphere set regions. Each dot represents a single microsphere, with the color change indicating increasing numbers of overlapping data points.. Microspheres are interrogated as they pass single file by a red and green laser.

Sequencing

Amplicon sequencing provides accurate, rapid and simple way of detection and identification of products. Sequencing allows us to accurately perform species identification of microbial pathogens. DNA sequences of the bacterial 16S rRNA gene can be determined in a high-throughput by capillary electrophoresis of varied size fluorescently labeled PCR products that differ by 1 base pair from about 10 base pairs in size up to 500. This high resolution separation gives 4 different fluorescent colors for the 4 different bases (A, C, T, & G) allowing one to tell the contiguous sequence of the 500 base pair PCR product. Comparison of this sequence with a large database (Genbank) allows one to identify the organism to species. Using the sequencing technique, one can also genotype organisms (hepatitis C virus), and evaluate mutations that may lead to resistance to certain medications such as antivirals in HIV.

Several newer sequencing technologies exist that provide higher throughput than traditional Sanger sequencing that was developed over 30 years ago. Pyrosequencing is a non-gel method which utilizes a real-time evaluation based on monitoring enzymatic inorganic pyrophosphate release by DNA polymerase. Light is produced and is proportional to the amount of pyrophosphate that is generated, and this is related to the nucleotides in the sequence. Nucleotides that are not used in the reaction are eliminated by apyrase before addition of the next nucleotide. Sequence data is quantitative and generated in real-time. Multiple commercial pyrosequencing systems have been used widely to rapidly identify screen for antibiotic drug resistance and identify particular microorganisms. A variant of this technology is multiplexed pyrosequencing which allows for the simultaneous extension of several primers hybridized to multiple target templates. This has been utilized extensively in the fields of medical genetics, cytogenetics and pharmacogenetics. A large scale variation in this technology is the Roche 454 deep sequencing method. It utilizes pyrosequencing without cloning bias and it can create 20 million bases (200,000 reads of 100 bp) of sequence data in a short time frame (hours).

Conclusions

In this chapter we have discussed the general methods for detecting organisms, there genotypes, and potential virulence factors and methods of antibiotic resistance. We have also mentioned the measurement of the patient's immune response and how it can be used in clinical diagnostics. Factors such as clinical utility, scientific accuracy, cost, ease of use all factor into the decision of what type of test to use. Most importantly one must evaluate what is best for the patient. We have seen that modern molecular methods are able to successfully and accurately diagnose infections in which traditional culture lacks sensitivity. Genome wide assessment is the future of microbiology and will be used to tailor treatment to specific patients in order to provide the best care available.

Acknowledgments

Figures 1B-C, 1E-F, and 3E-3G were kindly provided by Melissa Miller, Ph.D. Figure 10.6 was kindly provided by John Schmitz, Ph.D., Figure 10.5 was adapted from http://virology.org/topics.html. Figure 10.9 was kindly provided by Sherry Dunbar, Ph.D.

References

[1] Detrick B, Hamilton RG, Folds JD. *Manual of molecular and clinical laboratory immunology,* 7th ed. Washington, D.C.: ASM Press, c2006.

[2] Forbes BA, Sahm DF, Weissfeld AS, Bailey WR. *Bailey & Scott's diagnostic microbiology* 12th ed. St. Louis, Mo.: Elsevier Mosby, 2007.

[3] McPherson RA, Pincus MR, Henry JB. *Henry's clinical diagnosis and management by laboratory methods*. 21st ed. / [edited by] Richard AM et al. Philadelphia : Saunders Elsevier, c2007.

[4] Lennette EH, Lennette DA, Lennette ET. *American Public Health Association. Committee on Laboratory Standards and Practices. Diagnostic procedures for viral, rickettsial, and chlamydial infections.* 7th ed. Washington, DC: American Public Health Association, c1995.

[5] Specter S, Hodinka RL, Young SA. *Clinical Virology Manual* 4th ed. Washington, DC:ASM Press, c2009.

[6] Murray PR, Baron EJ, Jorgensen JH, Landry ML, Pfaller MA. *Manual of clinical microbiology*. 9th ed. Washington, D.C.: ASM Press, c2007.

[7] Persing DH, Tenover FC, Versalovic J, Tang YW, Unger ER, Relman DA, White TJ. *Molecular microbiology : diagnostic principles and practice.* Washington, DC : ASM Press, c2004.

[8] Tang YW, Stratton CW. *Advanced techniques in diagnostic microbiology* New York, NY: Springer, c2006.

[9] Leonard DG. *Molecular pathology in clinical practice. Infectious diseases* New York: Springer, c2009.

[10] Winn, WC, Koneman EW. *Koneman's color atlas and textbook of diagnostic microbiology*. 6th ed. Philadelphia: Lippincott Williams & Wilkins, c2006.

[11] Bennett JV, Jarvis WR. *Bennett & Brachman's hospital infections.* 5th ed. / editor, William R. Jarvis. Philadelphia : Wolters Kluwer Lippincott Williams & Wilkins, c2007.

[12] Dorak MT. *Real-time PCR*. New York, NY : Taylor & Francis, c2006.

[13] Mandell GL, Bennett JE. *Dolin, Raphael Mandell, Douglas, and Bennett's principles and practice of infectious diseases.* 7th ed. Philadelphia: Churchill Livingstone/Elsevier, c2010.

[14] Tang YW, Persing DH. Diagnostic Microbiology in *Encyclopedia of Microbiology*, pp 308-20, [edited by] Schaechter M. Oxford: Elsevier, c2009

Index

A

D

E

F

G

H

I

J

K

L

M

N

O

P

T

U

V

W

X

Y

Z